Reactive Oxygen Species in Plant Biology

Soumen Bhattacharjee

Reactive Oxygen Species in Plant Biology

Springer

Soumen Bhattacharjee
Department of Botany, UGC Centre for Advanced Study
The University of Burdwan
Burdwan, West Bengal, India

ISBN 978-81-322-3939-0 ISBN 978-81-322-3941-3 (eBook)
https://doi.org/10.1007/978-81-322-3941-3

Library of Congress Control Number: 2019934759

This Springer imprint is published by the registered company Springer Nature India Private Limited.
The registered company address is: 7th Floor, Vijaya Building, 17 Barakhamba Road, New Delhi 110 001, India

Dedicated to
My little Soham living in the peaceful abode of Holy Mother

Foreword

Reactive oxygen species (ROS) are partially reduced or excited forms that are generated by oxygen in cells. As by-products of aerobic metabolism, ROS are primarily formed in various cell organelles in plants including chloroplast, mitochondria, and peroxisome. The process of ROS production as a by-product of aerobic metabolism is coupled with ROS removal by cellular antioxidative mechanisms. It occurs constantly in cells to tightly regulate the endogenous ROS titer to prevent some of their potential toxic effects. ROS are found to regulate multiple processes in cells such as cellular redox levels, metabolism, development, stress signaling, systemic responses, and cell death in higher plants. The chemistry and physiology of the different ROS in plant cells exhibit features typical for second messengers with selective advantages. As signaling molecules, ROS are highly versatile due to their diverse properties that include different levels of reactivity, high rate of turnover, diverse sites of production, and potential to cross the biological membranes. The book *Reactive Oxygen Species in Plant Biology* covers the latest advances made in the field. It provides insights into ROS formation, sensing, detox scavenging, role in oxidative deterioration, and signaling associated with redox regulatory in plant. Therefore, this book is a valuable guide not only for students and teachers but also for researchers working in this field. The molecular language associated with ROS-mediated signal transduction leading to modulation of gene expression, which determines the stress acclamatory performance of plants, is an important content of this book. Importantly, the book provides detailed information on the current trends in redox proteomics and genomics to provide a complete understanding on the role of these redox players in cellular processes. The book is indeed a comprehensive one that covers the new developments in plant ROS biology.

Unique Features

1. The present book covers the basic niche area of plant redox biology for understanding the physicochemical basis of the formation of ROS, their properties, associated antioxidant defense mechanisms, oxidative damages, and implication of oxidative stress in plant cell.
2. The book also covers the basic information related to redox regulation of two important physiological processes, photosynthesis and senescence.

3. The book also provides information on the different aspects of redox signaling, where ROS may be projected as a central player, particularly under environmental stress and infection.
4. The information on the current trend in redox omic sciences is also accommodated to enable the readers to understand the application of plant redox biology in agriculture.

FNA, FNASc, FASc, JC Bose National Fellow
Professor, Department of Biotechnology
National Institute of Technology Durgapur
Durgapur, West Bengal, India

Sudip Chattopadhyay

Preface

Reactive oxygen species (ROS) most likely appeared on Earth together with the first atmospheric oxygen molecules about 2.4–3.8 billion years ago and have been a constant companion of aerobic life. In fact, the evolution of aerobic life on Earth occurred in the presence of ROS, and this fact should be kept in mind when we consider the roles that ROS play in the different biological systems. As toxic by-products of aerobic metabolism, ROS, are primarily formed in chloroplasts, mitochondria, peroxisomes, and apoplast and also at any other cellular locale. The process of ROS production as a by-product of aerobic metabolism, coupled with ROS removal by cellular antioxidative mechanisms, occurs constantly in cells to tightly regulate their endogenous titer and to prevent some of the potential toxic effects of ROS that could attack almost every important biomolecule. ROS were found to regulate development, differentiation, redox levels, stress signaling, interactions with other organisms, systemic responses, and cell death in higher plants. Further, the chemistry and physiology of different ROS in plant cell exhibit features typical for second messengers with selective advantages. As signaling molecules, ROS are highly versatile owing to their diverse properties that include different levels of reactivity, high rate of turnover, diverse sites of production, and potential to cross biological membranes. The origin of redox signal with antioxidants and ROS interacting at metabolic interface is a universal feature of plants. In fact, the unfavorable environmental cues trigger on specific ROS signaling, which are buffered by antioxidants through modulation of redox status and subsequent sensing mechanisms. As a consequence, several redox-sensitive signal transductions in different cellular locations got activated, which eventually play a vital role in plant performance under stress.

The aging and senescence processes of plants also witness loss of redox homeostasis. The tissue necrosis triggered by ROS under stress is a common feature of plant. A range of redox-sensitive enzyme system has been implicated in ROS interaction during programmed cell death (PCD). The elevated accumulation of ROS might cause nonspecific oxidative damages to every important class of biomolecules and cause oxidative damage and metabolic dysfunction. Similarly, one of the earliest events upon recognition of plant pathogen is also accumulation of ROS, which plants subsequently exploit as signaling agent for triggering defense processes. Chloroplast and peroxisome are two prime sources of ROS in plant cell, and their redox status plays significant role in determining the photosynthetic ability of plant.

Recently, ROS have been recognized as a central component of complex signaling network under environmental stress. The participation of ROS in redox signaling associated with stress acclimation also demands coordinated regulation of antioxidative defense in plant. The network of ROS-sensitive redox-regulatory genes is highly dynamic and redundant that encodes redox-sensitive proteins, ROS-scavenging and ROS-producing molecules. ROS-induced retrograde signaling is extremely important in regulating metabolic fluxes, optimizing metabolic and cellular functions. Apart from its bona fide role in signaling cascades, ROS often complement, synergize, and antagonize other growth regulatory circuits through cross talking with other signaling molecules. Therefore, the significance of ROS in plant life is beyond question, and its impact depends not only on ROS metabolism and functioning of its network but also on capacity of the cell in maintenance of redox status. So, to understand and explore oxidative stress, accurate assessment of redox status of the cell is required, which demands sensitive and robust assay for determination and quantification of ROS in plant cell as well as characterization of redox changes at metabolomic, proteomic, and genomic level.

The aim of this book is to provide in-depth information on ROS biology covering overall aspects of chemistry, generation, their role and regulation of metabolism, signaling under stress, and management for crop improvement. The book contains nine chapters that follow a sequence of gradual unfolding of impact of ROS in plant biology. It is indeed an effort toward a timely contribution of a topic that is of great importance to the understanding of physiology of plant and also future food security. Chapter 1 describes our current perception on the origin and implication of ROS and oxidative stress. Chapter 2 explores the relationship between ROS and antioxidants in plant cell. It provides information on how antioxidants provide essential information on cellular redox state that influences the expression of stress acclamatory genes in plant. The complex regulatory mechanisms function at molecular level to coordinate antioxidant response required for maintenance of redox status of plant cell. Chapter 3 aims at understanding the role of ROS in aging and senescence. Chapter 4 unfolds the role of ROS in oxidative modification of cellular components and their subsequent physiological significance. Chapter 5 provides a comprehensive overview of the role of ROS in photosynthesis. Chapter 6 deals with ROS-associated stress tolerance and signaling with an updated concept of ROS signaling in maintaining energy and metabolic fluxes. And finally, Chap. 7 overviews the prospects of proteomic and genomic approaches in exploring oxidative stress in plants.

Hopefully, this book will serve as a major source of information for PG students, research scholars, and professionals in the field in investigating the role of ROS in plant biology. It will definitely be of interest to a wide range of plant scientists, including plant biochemists, molecular biologists, plant physiologists, geneticists, breeder, agronomists, and biotechnologists, who have concern about the plant performance under stress and also cell signaling mechanisms.

Burdwan, West Bengal, India Soumen Bhattacharjee

Acknowledgment

As an author, I extend my thanks to Raman Shukla, Sr. Editorial Assistant (Life Sciences and Biomedicine), and RaagaiPriya ChandraSekaran, Project Coordinator (Books), of Springer Nature, India, who enabled me to initiate this book project and exhibited their long patience for this book to materialize. Special thanks are also extended to Springer Nature, India, for publishing this book. I would like to express my appreciation and thanks to my research group of Plant Redox Biology of Plant Physiology and Biochemistry Laboratory, Department of Botany, University of Burdwan, West Bengal, India (Dr. Manashi Aditya, Dr. Ananya Chakrabarty, Nabanita Banik, Nivedita Dey, Durga Kora, Debasmita Sen, Utpal Krishna Roy, Sarmistha Ghosh, Ananya Dey, Babita Pal, Agnideepa Kar) for their kind assistance in preparing the manuscript. My sincere gratitude to my teacher Prof. Asok Kumar Mukherjee for his kind blessings and advice is sincerely acknowledged. Research grants of different agencies like UGC, DST (WB), UGC (CAS), CSIR, etc., which have enabled me to work in plant redox biology during the last two decades, are gratefully acknowledged. I must also record my special appreciation for my wife Lily Bhattacharjee, who has always been my source of strength and inspiration. Lastly, I am always indebted to Swami Vivekananda, Holy Mother, and Ramakrishna Paramahansa for inner spiritual inspiration.

Contents

1 ROS and Oxidative Stress: Origin and Implication 1
1.1 Introduction 2
1.2 Chemistry of ROS 4
1.2.1 Superoxide Radical 5
1.2.2 Perhydroxy Radical 6
1.2.3 Hydrogen Peroxide 6
1.2.4 Singlet Oxygen 9
1.2.5 Hydroxyl Radical 9
1.3 Generation of $O_2^{\cdot-}$, H_2O_2, and $OH^{\cdot}$ and Other ROS in Plant Cells 10
1.3.1 Chloroplast and Peroxisome-Associated Generation of ROS 10
1.3.2 Mitochondrial Generation of ROS 13
1.3.3 Microsomal and Apoplastic Generation of ROS 14
1.3.4 NADPH Oxidase (RBOH)-Induced Formation of ROS 15
1.3.5 Cell Wall and Other Enzymatic Sources of ROS 15
1.4 Membrane Lipid Peroxidation: A Prospective Source of Oxyfree Radicals in Plant Cell 16
1.5 Abiotic Stress and Production of ROS 18
1.5.1 Generation of ROS Under Biotic Stress 21
1.6 ROS Wave 23
1.7 ROS Generation During Senescence 24
1.8 Conclusion and Perspective 24
References 25

2 ROS and Antioxidants: Relationship in Green Cells 33
2.1 Introduction 34
2.2 Antioxidant Defense Machinery of Plant Cell 35
2.2.1 Enzymatic Antioxidants 35
2.2.2 Nonenzymatic Antioxidants and Free Radical Quenchers 41
2.3 Regulation of Antioxidative Defense System in Plants as a Mechanism to Combat Environmental Stress 44

2.4 ROS-Antioxidant Interaction at Metabolic Interface Determines Redox Signaling Under Environmental Stress 49
2.5 Antioxidant and Redox Sensing Under Environmental Stress 53
2.6 Conclusion 54
References 55

3 ROS in Aging and Senescence 65
3.1 Introduction 66
3.2 Implication of ROS in Plant Senescence 67
3.3 Implication of ROS and Oxidative Stress in Programmed Cell Death of Plant 69
3.3.1 Tilting ROS Homeostasis to Induce Signaling During Senescence 71
3.4 ROS Signaling Associated with Programmed Cell Death of Plants 72
3.5 ROS Signal Communication During Cell Death 74
3.6 Conclusion and Perspective 76
References 76

4 ROS and Oxidative Modification of Cellular Components 81
4.1 Introduction 82
4.2 Oxidative Modification of Membrane Lipid 84
4.2.1 Mechanism of Membrane Lipid Peroxidation (MLPO) 85
4.2.2 Enzymatic and Nonenzymatic Membrane Lipid Peroxidation (MLPO) Works in Tandem in Plant Cell with Physiological Relevance 86
4.2.3 MLPO Serves Other Physiological Purpose Apart from Oxidative Damage 89
4.2.4 Some Secondary Products of MLPO Work as Reactive Lipid Species 90
4.3 Oxidative Modification to Protein 92
4.3.1 Mechanism of Oxidative Modification of Cellular Proteins 93
4.3.2 Abiotic Stress and Oxidative Modification of Proteins 94
4.4 Cross Talk Between ROS-Mediated Protein Oxidation and Lipid Peroxidation 95
4.5 Assessment of Products of MLPO and PO as Sensitive Redox Biomarkers for the Evaluation of Impact of Environmental Stress 95
4.6 Oxidative Damage to Nucleic Acids 98
4.7 Conclusion and Perspective 99
References 100

5 ROS and Regulation of Photosynthesis 107
5.1 Introduction 108
5.2 Origin of Photooxidative Stress 109

5.3 ROS-Antioxidant Interaction and Regulation of Redox Status of Chloroplast 112
5.4 An Elaborate Antioxidant Network Works at Chloroplast for the Management of Oxidative Stress and the Origin of Redox Signal for Combating Oxidative Stress 114
5.4.1 Ascorbate-Glutathione (ASA-GSH) Cycle or Halliwell-Asada Pathway 114
5.4.2 Water-Water Cycle or Asada Pathway 116
5.4.3 ROS-Induced Photoinhibition and Its Significance in the Maintenance of Redox Homeostasis of Green Cell Under Stress 117
5.4.4 Xanthophyll Cycle and Its Role in Photoprotection 119
5.5 Origin of Chloroplast Redox Signal and Control of Photosynthesis 121
5.6 Conclusion and Perspective 122
References 123

6 ROS: Central Component of Signaling Network in Plant Cell 127
6.1 Introduction 128
6.2 ROS as Signaling Molecule with Specific Advantages 131
6.2.1 ROS (H_2O_2) Transport 131
6.3 ROS Signal with Specific Targets in Plant Cell 132
6.3.1 Redox Sensors of Plant Cell 134
6.3.2 Some Recognized H_2O_2 Sensors of Plant Cell (OxyR, Hsf1, PerR, PTPs, HSFs, YAP-1) 136
6.3.3 Role of Redox-Sensitive Proteins (RSPs) 138
6.3.4 Redox Signaling and Regulation of Ca^{2+} Homeostasis and MAPK Cascades 139
6.4 H_2O_2-Responsive Transcription Factors 140
6.5 Redox Signaling, miRNAs, and Gene Expression 142
6.6 ROS Cross Talk with Other Signaling Pathway Under Environmental Stress 143
6.7 Lipid Peroxidation Products as "Biological Signals" 146
6.8 Conclusion and Perspective 147
References 148

7 Exploring Oxidative Stress in Plants: Proteomic and Genomic Approaches 155
7.1 Introduction 156
7.2 Different Proteomic Approaches Used for the Identification of ROS-Modified Proteins 157
7.2.1 Proteomic Approach for Molecular Characterization of Oxidized Protein Residues 159
7.2.2 Two-Dimensional Electrophoreses Coupled with MALDI TOF MS/MS 162

7.2.3 Strategies to Study Posttranslational Modifications in Plants in Response to Stress 164
7.2.4 Strategies to Study S-Glutathiolation and N-Nitrosylation of Proteins........ 167
7.2.5 Proteomic Approach to Identify Carbonylation of Proteins Under Oxidative Stress 168
7.3 Genomic Approach for Understanding Oxidative Stress Responses in Plants 169
7.3.1 Transcriptomic Study........ 170
7.3.2 Genome-Wide Association Studies (GWAS) and Quantitative Real-Time Reverse-Transcriptase PCR Approach for Understanding Oxidative Stress Responses in Plants 171
7.4 Genetic Engineering for Oxidative Stress Tolerance........ 172
7.4.1 Transgenesis for Oxidative Stress-Tolerant Plants Targeting Antioxidant Genes 174
7.4.2 Transgenesis for Oxidative Stress-Tolerant Plants Targeting Sulfur Metabolism Pathway Genes 174
7.4.3 Transgenesis for Oxidative Stress-Tolerant Plants Targeting Transcription Factors and Signal Transduction Pathways 175
7.5 Conclusions and Future Perspectives........ 177
References........ 177

About the Author

Dr. Soumen Bhattacharjee, presently Professor and Program Coordinator, UGC Centre for Advanced Studies, Department of Botany, the University of Burdwan, West Bengal, did his Masters in Botany and Ph.D. on Abiotic Stress Physiology of Plants from the University of Burdwan, West Bengal, India. Later, he started his teaching career in the year 1995 as a Faculty in the Department of Botany of Delhi University's constituent college situated in Bhutan (SAARC country). After serving almost two and a half years, he joined West Bengal Education Service and worked mainly in the Postgraduate Department of Botany, Hooghly Mohsin College, West Bengal. In the year 2007, Dr. Bhattacharjee got selected by the Agricultural Scientists Recruitment Board, Indian Council of Agricultural Research (ICAR), as Senior Scientist and joined Vivekananda Institute of Hill Agriculture, Almora, India. In 2013, he joined the University of Burdwan as Professor. In his teaching career, Dr. Bhattacharjee had the privilege of teaching PG and UG courses at the University of Burdwan, University of Kalyani, and University of Delhi.

The research interest of Dr. Bhattacharjee centers around plant redox biology, particularly understanding the relationship between oxidative stress, growth and yield potential of plant, role of reactive oxygen species (ROS) signaling in stress acclimation, characterization of redox-regulatory mechanism during germination of rice under abiotic stress, understanding physiological basis of antioxidant accumulation in underutilized medicinal plants, etc. He has published 90 publications (58 research papers in international peer-reviewed journals, 2 edited books in Springer International, 9 edited volumes, and 21 articles in proceedings of the national and international conferences). The author at present possesses 26 i10-index and 16 h-index with more than 1200 citations of his research work. At present, a group of nine doctoral students are actively working under his guidance in the domain of plant redox biology with different national and international fellowships.

ROS and Oxidative Stress: Origin and Implication

1

Abstract

Molecular oxygen (O_2) is the primary cellular electron acceptor in aerobic respiration that serves fundamental roles in membrane-linked ATP formation and other fundamental cellular and metabolic functions. But, as an untoward but inescapable consequence of different metabolic events in oxygen-saturated cellular environment, reactive oxygen species (ROS) are incessantly generated by partial or incomplete reduction of molecular oxygen. In plants, ROS are continuously generated as oxidation – reduction cascades of different metabolism located in different cellular compartments and as by-product of various metabolic events. The most important ROS include superoxide ($O_2^{\cdot-}$), perhydroxy radical ($HO_2^{\cdot}$), hydrogen peroxide (H_2O_2), hydroxy radical ($OH^{\cdot}$), and singlet oxygen ($^1O_2$). The other secondary oxidative products like alkoxy radical ($RO^{\cdot}$), peroxy radical ($ROO^{\cdot}$), organic hydroperoxide (ROOH), excited carbonyl ($RO^{\cdot}$), etc. are also produced in plant cells. Though ROS is generated under natural conditions, their productions are augmented under the exposure of unfavorable environmental cues and natural course of senescence. Major sources of ROS in plant cell encompass spilling of electrons during photosynthetic and respiratory electron transport, decompartmentalization of transition metal ions, and also various biological redox reactions. In fact, the redox cascades of chloroplast, peroxisome, and mitochondria of green cells not only determine the driving forces for metabolism but also recognized as the prime source of ROS. Lipid peroxidation, which is known to produce ROS like alkoxy, peroxy radicals as well as singlet oxygen, is also considered as bona fide source of ROS in plant cells. In plants, apoplastic enzyme respiratory burst oxidase homologs (RBOHs) or NADPH oxidases play a major role in originating ROS wave through the other network of ROS production as well. The ROS wave, which is a consequence of perception of unfavorable environmental cues should be integrated with additional metabolic/signaling pathways to enable rapid systemic acclimation of plants. However, an elaborate and efficient antioxidative defense system, comprising a variety of antioxidant molecules, quenchers, and enzymes, determines the ROS turnover and hence the steady-state level of ROS

S. Bhattacharjee, *Reactive Oxygen Species in Plant Biology*,
https://doi.org/10.1007/978-81-322-3941-3_1

and the redox status of the cell. Plants are equipped with those defense systems not only to combat enhanced level of ROS but also to tightly regulate the endogenous concentration necessary for controlling various events of Plant Biology. However, the decontrolled level of ROS generation, if remaining unabated may cause a solemn threat to or cause oxidative deterioration and in extreme cases the death of plant cells. The present chapter describes the physicochemical basis of the production of ROS, under normal and unfavorable environmental conditions, and senescence, with an added effort to understand their implication associated with those situations.

Keywords
Oxyfree radicals · Oxidative stress · Antioxidative defense · Environmental stress · ROS wave · Environmental stress

1.1 Introduction

Earth is the only planet that sustains aerobic life in our solar system primarily because it contains molecular oxygen in its atmosphere. However, after its origin, about 4.6 billion years back, its atmosphere was primarily reducing and essentially devoid of molecular oxygen. The traces of oxygen in the primordial earth were probably due to the photolytic (sun's ionizing radiation) dissociation of water. The present oxygen concentration (~21% by volume) of the earth is the outcome of oxygenic photosynthesis (Halliwell 2006). It has been estimated that molecular oxygen present in the hydrosphere as water is only 37 Erda moles out of nearly about 410×10^3 Erda moles of total oxygen (Emol = 10^8 mol). Atmospheric and hydrospheric molecular oxygen (37 Emol and 0.4 Emol, respectively) undergoes continuous turnover, with the total oxygen exchange rate at ~15 × 103 Emol/10^6 years. And it is the aerobic life which is responsible for the major oxygen turnover, with the process photosynthesis being the chief input into the oxygen reservoir and respiration the major output (Eltsner 1987; Scandalios 2005).

The formation of atmospheric oxygen may well stand for one of the surprising event of planet's history. The molecular oxygen was introduced approximately 2.7 billion years ago in our environment by the evolution of O_2-evolving photosynthetic organisms, and there onward reactive oxygen species (ROS) have been the undesirable and inseparable companions of aerobic life (Mittler 2017; Halliwell 2006). The atmospheric oxygen initially produced by oxygenic photosynthesis on Earth was almost immediately converted into ROS or oxyfree radicals due to the occurrence of the high levels of soluble Fe^{2+} in the primordial oceans. The ubiquitous distribution of antioxidative enzyme superoxide dismutase (SOD), which has evolved even before the evolution of eubacteria from archaea, strongly supports this possibility (Mittler 2017). Therefore, the evolution and diversification of aerobic life on Earth primarily took place in the presence of ROS, and this significant fact should be considered seriously when we judiciously adjudge the roles that ROS play in different living systems.

The O_2 molecule has two unpaired electrons in their outermost orbitals that have the same spin quantum number. In fact, this spin restriction favors O_2 to accept its electrons in its outermost orbital one at a time, leading to the generation of the transient highly reactive intermediates, the so-called ROS, which subsequently became a strong oxidant. In fact, this class of reactive species, which sustains aerobic life, can act as a deadly contaminant in mildly reducing cellular atmosphere, particularly when the redox homeostasis of the cell is disrupted (Foyer and Shigeoka 2011; Das and Roychoudhury 2014; Levine 1999; Miller et al. 2008). Therefore, although aerobic metabolism is energetically efficient, the oxygen-saturated condition in cellular environment possesses a constant oxidative threat to cellular metabolites, structures, and processes. So, the evolution of metabolic processes, where O_2 is associated in one way or another, such as photosynthesis, aerobic respiration, and photorespiration, undoubtedly has the probability of formation of ROS. An inevitable result of chloroplastic, mitochondrial, and plasma membrane-linked electron flow is the spilling of electrons onto molecular oxygen, particularly under redox imbalanced situation, with the resultant generation of reactive oxygen species (ROS) (Alscher et al. 1997; Arora et al. 2002; Bhattacharjee 2005; Moller et al. 2007; Miller et al. 2008, 2010). The imposition of environmental stresses can further upregulate the production of ROS (Arora et al. 2002; Bhattacharjee 2005; Alscher and Hess 1993; Fridovich 1995; Das and Roychoudhury 2014). ROS are also produced during normal metabolic events (Alscher and Hess 1993; Hammond-Kosack and Jones 2000). About 1–2 % of O_2 consumed by plants is diverted to produce ROS in various subcellular loci under normal ambient condition (Eltsner 1987; Del Rio et al. 1992; Das and Roychoudhury 2014).

The redox status and its homeostasis in plants is regulated primarily by two arms of the antioxidant defense machinery, the enzymatic components and the nonenzymatic low molecular compounds. In fact, in a normal plant cell, antioxidative defense systems are in constant vigil to tightly regulate the endogenous titer of ROS. Exposure of plants to unfavorable environmental stress results in oxidative stresses that shift the balance in favor of ROS or prooxidants (Asada and Takahashi 1987; Miller et al. 2008, 2010). ROS, namely, $O_2^{\cdot -}$, H_2O_2, and $OH^{\cdot}$, can be formed during the one electron-mediated reduction of O_2 to H_2O (Fig. 1.1). ROS capable of causing oxidative damage include superoxide ($O_2^{\cdot -}$), perhydroxy radical ($HO_2^{\cdot}$), hydrogen peroxide (H_2O_2), hydroxy radical (OH), singlet oxygen ($^1O_2$), peroxy radical ($ROO^{\cdot}$), alkoxy radical ($RO^{\cdot}$), organic hydroperoxide (ROOH), excited carbonyl ($RO^{\cdot}$), etc.

In fact, O_2 is required in aerobic respiration to produce the energy needed for maintaining steady-state conditions. During normal cellular metabolism, ground-state O_2 is reduced to H_2O via $O_2^{\cdot -}$, $HO_2^{\cdot}$, H_2O_2, and $OH^{\cdot}$. Thermodynamically, ROS is produced from O_2 either by electron transfer or energy transfer reactions. Though initially, the reaction cascade requires an energy input, the subsequent steps are found to be exothermic and can occur spontaneously. Acceptance of excess energy (photochemical) by O_2 can additionally lead to the formation of $^1O_2$, a highly toxic ROS, by reverting the movement of electron flow. So, any condition which has a tendency to disrupt the redox homeostasis of the cell, where the redox steady state is altered in the direction of prooxidants, primarily due to the overproduction of ROS or incapability of antioxidative

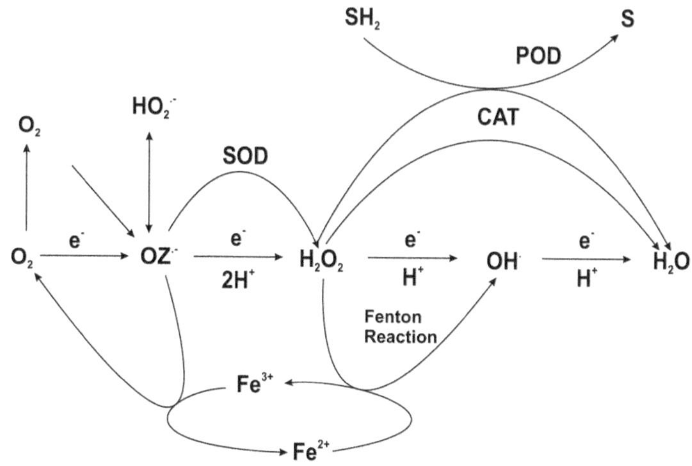

Fig. 1.1 Generation and interconversion of ROS derived from O_2. The ground state of molecular oxygen can be activated by excess photochemical energy or photoexcitation, reverting the spin of one of the unpaired electrons to form singlet oxygen ($^1O_2$). One electron reduction leads to the formation of superoxide ($O_2^{\cdot-}$) radical. Superoxide exists in equilibrium with conjugate acid, hydroperoxy radical (HO_2^-). Subsequent one electron-mediated reduction then produces hydrogen peroxide (H_2O_2), hydroxyl radical ($OH^\cdot$), and finally water (H_2O). Metal ions that are mainly present in cells in oxidized form (Fe^{3+}) are reduced in the presence of $O_2^{\cdot-}$ and consequently may catalyze the conversion of H_2O_2 to OH. by Fenton or Haber-Weiss reaction. Enzymes superoxide dismutases (SOD), catalases (CAT), and peroxidases (POD) reduce ROS. POD require a reducing substrate SH2 for the reduction

defense system to scavenge them efficiently or the both, is recognized as oxidative stress. The manifestation of this state of cell, i.e., oxidative stress, which leads to the rapid generation and subsequent accumulation of ROS, ranges from metabolic dysfunction to loss of cellular architecture and genomic lesions (Bhattacharjee 1998, 2005; Bhattacharjee and Mukherjee 2004; Wiseman and Halliwell 1996; Moller et al. 2007).

The enzymatic and nonenzymatic components that buffer or tight regulate the redox homeostasis of the plant cell include superoxide dismutase (SOD), ascorbate peroxidase (APX), guaiacol peroxidase (GPX), glutathione-S-transferase (GST), and catalase (CAT) and the nonenzymatic low molecular compounds like ascorbic acid (AA), reduced glutathione (GSH), other thiol compounds, α-tocopherol, carotenoids, phenolics, flavonoids, proline, etc. (Miller et al. 2010; Gill et al. 2011). The ubiquitous nature of both the components efficiently maintain and support the antioxidant defense machinery necessary for the detoxification of ROS for survival of cell (Bhattacharjee 2005; Gill et al. 2011).

1.2 Chemistry of ROS

The different chemotypes of ROS that plant cells can produce both under normal and stressed situations and are capable of causing oxidative damage include superoxide ($O_2^{\cdot-}$), perhydroxy radical ($HO_2^\cdot$), hydrogen peroxide (H_2O_2), hydroxy radical

(OH·), alkoxy radical (RO·), peroxy radical (ROO·), organic hydroperoxide (ROOH), singlet oxygen ($^1O_2$), excited carbonyl (RO·), etc. O_2 in its ground state is totally a harmless molecule and possesses two unpaired electrons with parallel spin which offer its paramagnetic property and, hence, improbable to instigate reactions with biomolecules in cellular environment, unless it is energetically activated (Apel and Hirt 2004). This activation of O_2 may occur either by absorption of sufficient energy to reverse the spin on one of the unpaired electrons or stepwise addition of single electron (monovalent reduction) (Fig. 1.1). In the former case, $^1O_2$ is formed, whereas in latter, O_2•–, H_2O_2, and •OH are formed sequentially (Fig. 1.1).

1.2.1 Superoxide Radical

When oxygen takes up one electron as leaks from photosynthetic electron transport (PET) and mitochondrial ETC, superoxide (O2$^{-\cdot}$) is formed. Its main production sites are PSI of Z-scheme of photosynthesis (from ferredoxin) and the internal mitochondrial membrane (from NADH ubiquinone reductase and ubiquinone cytochrome c reductase). Subsequently, this radical species through one electron-mediated reduction forms hydrogen peroxide (H_2O_2). The production of superoxide radicals at the membrane level, facing apoplast, is mediated by NADPH oxidase (RBOH) and primarily initiated during oxidative burst with phagocytic functions and contributes to their bactericidal action (Fridovich 1986). In plants, NADPH oxidases or respiratory burst oxidase homologs (RBOHs) play a key role in the network of ROS production, commonly called as ROS wave, initiating with the generation of superoxide (Suzuki et al. 2011).

$O_2^{\cdot-}$ are produced continuously during pseudocyclic electron flow of photosynthetic Z-scheme in the chloroplasts by partial reduction of O_2 molecules or energy transfer to them. PSI localized in thylakoid membrane is found to be the major site of $O_2^{\cdot-}$ production during photosynthesis. The $O_2^{\cdot-}$ thus produced upon reduction of O_2 during PET is known as pseudocyclic pathway of chloroplasts. The production of ROS like $O_2^{\cdot-}$ is an inevitable consequence of aerobic respiration as well. When the terminal cytochrome c oxidase or the alternative oxidase react with O_2, four electrons are transferred sequentially and H_2O is formed. However, under oxidative stress or loss of redox homeostasis, O_2 can react with other ETC components. Here, only one electron is transferred instead of four, resulting in the formation of $O_2^{\cdot-}$, a moderately reactive ROS. The generation of $O_2^{\cdot-}$ may further trigger the production of more active ROS like OH·, which may cause initiate membrane damage by peroxidation to membrane lipids. It has been noted that $O_2^{\cdot-}$ can undergo protonation to give $HO_2^{\cdot-}$, a strong oxidant, in negatively charged surfaces of the membrane, which may directly attack the membrane lipid-associated PUFA. Furthermore, $O_2^{\cdot-}$ can also reduce Fe^{3+} iron to yield a reduced form of iron (Fe^{2+} iron), which got the capacity to reduce H_2O_2, produced as a result of SOD-led dismutation reaction of $O_2^{\cdot-}$ to $OH^{\cdot-}$. These cumulative reactions in cellular environment through which $O_2^{\cdot-}$ and H_2O_2 and iron rapidly generate the most toxic ROS $OH^{\cdot-}$ is called the Haber and Weiss reaction, whereas the terminal final step which involves the oxidation of Fe^{2+} by H_2O_2 is referred to as the Fenton's reaction.

$$O^{\bullet-2} + Fe^{3+} \rightarrow {}^1O_2 + Fe^{2+}$$

$$2O^{\bullet-2} + 2H^+ \xrightarrow{SOD} O_2 + H_2O_2$$

$$Fe^{2+} + H_2O_2 \rightarrow Fe^{3+} + OH^- + OH\bullet \left(\text{Fenton reaction}\right)$$

$O_2^{\cdot-}$ is a moderately reactive, short-lived ROS with a half-life of approximately 2–4 μs (Table 1.1) (Dat et al. 2000). $O_2^{\cdot-}$ is impermeable to biological membranes as it gets dismutated to H_2O_2. However, $O_2^{\cdot-}$ is extremely reactive in hydrophobic environment such as interior of membrane or multimeric protein

1.2.2 Perhydroxy Radical

The protonated form of $O_2^{-\cdot}$, $HO_2^{\cdot}$ is more reactive than superoxide itself, but in plant cells at physiological pH a very small proportion of $O_2^{-\cdot}$ would be in this form (Eltsner 1987). However superoxide can dismutate to form H_2O_2. And much more reactive OH· can be formed from $O_2^{-\cdot}$ and H_2O_2 through Fe-catalyzed Haber-Weiss reaction.

1.2.3 Hydrogen Peroxide

The univalent reduction or one electron-mediated reduction of $O_2^{\cdot-}$ produces H_2O_2 which is moderately reactive and has relatively longer half-life (1 ms) as compared to other ROS. It has been well established that excess titer of H_2O_2 in the plant cells primarily leads to the occurrence of oxidative stress. H_2O_2 may inactivate enzymes and damage membrane by directly oxidizing their thiol groups and stimulating lipid peroxidation, respectively (Bowler et al. 1994; Buchanon and Balmer 2005). Therefore, H_2O_2, being moderately reactive (Table 1.1) and with relatively longer half-life (1 ms), can diffuse comparatively more distances from its site of production (Vanova et al. 2002). Since H_2O_2 has no unpaired electrons, it cannot be called as free radical, and unlike other oxyfree radicals, it is permeable to membranes and consequently can evoke their function far from the site of its formation, either oxidative damage or cell signaling. Because of these properties, i.e., its ability to diffuse through aquaporins in the membranes and migration capacity over larger distances within the cell and its relatively more half-life compared to other ROS, it has received serious attention as a "signal molecule" involved in the regulation of specific biological and developmental processes and triggering tolerance against various environmental stresses. All forms of membrane-linked electron flows associated with ATP formation, i.e., electron transport chain (ETC) of chloroplast, mitochondria, endoplasmic reticulum, and plasma membrane, along with metabolic cascades like β-oxidation of fatty acid and photorespiration are major sources of H_2O_2 generation in plant cells. Other reactions involving NADPH oxidase as well as xanthine oxidase (XOD) also contribute to H_2O_2 production in plants.

Table 1.1 Components of ROS and the antioxidative defense system of redox network in plant cell

[A]

ROS	Half life	Migration capacity	Subcellular location	Interacts with			References
				DNA	Protein	Lipid, carbohydrate	
Superoxide (O_2)	1–4 μs	30 nm	Membranes, Chloroplast (Mehler Reaction), Mitochondria	No	Yes (Fe-centre)	Hardly	Halliwell and Gutteridge (1984), Bhattacharjee (2012), Miller et al. (2010)
Hydrogen peroxide (H_2O_2)	1 ms	1 μm	Membranes, Chloroplast, Mitochondria, Peroxysome	No	Yes (Cysteine)	Hardly	Halliwell and Gutteridge (1984), Bhattacharjee (2012), Foyer (1997)
Hydroxyl radical (OH·)	1 μs	1 nm	Chloroplast, Membranes, Mitochondria	Rapidly	Rapidly	Rapidly	Halliwell and Gutteridge (1984), Bhattacharjee (2012)
Singlet oxygen ($^1O_2$)	1–4 μs	30 nm	Chloroplast, Membranes, Mitochondria	Yes (Guanine)	Trp, His, Tyr, Met, Cys.	PUFA	Halliwell and Gutteridge (1984), Bhattacharjee (2012), Foyer (1997)
Alkoxy radicals (RO·)	?	lnm	Membrane lipid peroxy dation	No	Yes	PUFA	Halliwell and Gutteridge (1984), Bhattacharjee (2012), Foyer (1997)
Peroxy radicals (ROO·)	?	lnm	Membrane lipid peroxy dation	No	Yes	PUFA	Halliwell and Gutteridge (1984), Foyer (1997)

(continued)

Table 1.1 (continued)

[B]

Antioxidant defense System	Interacts/Removes (product)	Sub-cellular location	References
Superoxide dismutase	$O_2^{\cdot-}$ (H_2O_2)	Chl, Cyt, Mit, Per	Chowdhury et al. (2016), Halliwell and Gutteridge (1984), Bhattacharjee (2012), Foyer (1997), Mittler (2017)
Catalase	H_2O_2 (H_2O)	Mit?, Per	Foyer (1997), Winston (1990)
Peroxidases	H_2O_2 (H_2O)	Many locations	Eltsner (1982)
Ascorbate/glutathione cycle	H_2O_2 (H_2O)	Chl, Cyt?, Mit, Per	Alscher and Hess (1993)
Glutathione-s-transferase	ROO	Nuel, Cyt	Alscher and Hess (1993)
Halliwell-Asada Path way	0_2^{-} (H_2O_2), H_2O_2 (H_2O)	Chi.	Halliwell and Gutteridge (1984), Foyer (1997)
Glutathione peroxidases	H_2O_2 (H_2O)	Chl, Cyt, ER, Mit	Creissen et al (1999)
	Lipid hydroperoxides		
	Other hydroperoxides		
Peroxiredoxin system	H_2O_2 (H_2O)	Chl, Cyt, Mit, Nuel	Foyer (1997), Rouhier et al. (2008)
	Alkyl hydroperoxides		
	Peroxini trite		
Thioredoxin system (Regulating SH/S=S ratio)	H_2O_2 (H_2O)	Chl, Cyt, Mit	Rouhier et al. (2008)
Glutaredoxin system (Regulating SH/S=S ratio)	H_2O_2 (H_2O)	Chl, Cyt, Mit, Sec	Foyer and Noctor (2003)
	Hydroperoxides		
Carotenes and tocopherol	$O_2^{\cdot-}$ (O_2)	Chl	Davison et al. (2002)

[A] The important ROS in plant tissues and their basic properties [Half life – in Biological system; Migration capacity – Distance traveled in one half life time if the diffusion co-efficient is assumed to be $10^{-9}m^{-2}s^{-1}$]. [B] Antioxidant mechanisms that scavenge and modulate the level of ROS (through ROS-removal mechanisms) in plant cell

Chl chloroplasts, *Cyt* cytosol, *ER* endoplasmic reticulum, *Mit* mitochondria, *Nuel* nucleus, *Per* peroxisomes, *Sec* secretory pathway, *SH/S=S*, Sulphydryl/Disulfide ratio

1.2.4 Singlet Oxygen

This chemotype of oxygen is not a true radical but is an extremely important ROS in reactions related to environmental stress-induced oxidative damages. The occurrence of transition metals contributes to significant enhancement of the production of singlet oxygen, as well as anion superoxide in the cellular environment and may cause lipid peroxidation. In ground state, the unpaired electrons in the of oxygen have parallel spin. Absorption of sufficient energy (may be from photoexcited antenna pigments) reverses the spin of one of its unpaired electrons leading to its conversion to singlet state in which the two outermost orbital electrons have opposite spin. This activation subsequently overcomes the spin restriction property of O_2, and $^1O_2$ can consequently participate in reactions involving the simultaneous transfer of two electrons (Apel and Hirt 2004). In presence of excess photochemical energy, highly reactive $^1O_2$ can be produced via triplet chlorophyll (Chl*) formation in the antenna system and in the reaction center of photosystem II (Krieger-Liszkay 2005). ^{1}O2 may be generated during the process lipid peroxidation as well and thus may cause the production of other peroxide molecules (hydroperoxide). $^1O_2$ when formed can either transfer its excitation energy to other biological molecules they encounter or cause lipid peroxidation, thus forming secondary products endoperoxides or hydroperoxides (Halliwell and Gutteridge 1999). The half-life of $^1O_2$ is very small and can stay for nearly 4 μs and 100 μs in water and polar solvents, respectively (Foyer and Harbinson 1994). Singlet oxygen is also very toxic, and its significance has been realized only recently, due to the development of methods for its generation and detection (Halliwell and Gutteridge 1999). The migration capacity of $^1O_2$ is found to be several hundred nanometers (nm) in cellular system.

1.2.5 Hydroxyl Radical

It is the most potent and toxic ROS in cellular system. H_2O_2 produced primarily by SOD can react with Fe^{2+}, produced by Haber-Weiss reaction to generate OH$^\cdot$. This reaction is called as Fenton reaction:

$$Fe^{2+} + H_2O_2 \rightarrow Fe^{3+} + OH + OH^-$$

It is so toxic that it can cause damage to different cellular components by lipid peroxidation (LPO), protein damage, DNA lesion, etc. Since there is no existing enzymatic system to scavenge this, and being a very strong, it exerts maximum potent toxic effect among all the ROS. There are evidences that excess accumulation of OH$^\bullet$ causes the cellular death (Pinto et al. 2003). Moreover, this iron-catalyzed decomposition of hydrogen peroxide is considered the most prevalent reaction in plants which causes deleterious lipid peroxidation, associated with membrane damage.

In addition, other ROS like peroxy and alkoxy radicals formed as intermediates of membrane lipid peroxidation and are also very toxic and poses threat to several biomolecules.

1.3 Generation of $O_2^{\cdot-}$, H_2O_2, and $OH^{\cdot}$ and Other ROS in Plant Cells

ROS can be produced effectively from almost every cellular compartment. ROS are produced with varying magnitude at several locations including chloroplasts, peroxisomes, mitochondria, plasma membranes, apoplast, endoplasmic reticulum, and cytosol, depending on the environmental and developmental condition (Fig. 1.2). Basically the prime process of ROS formation in plant cell, both under normal and environmental odds, is the spilling of electrons onto O_2 from the electron transport chains of chloroplasts, mitochondria, and plasma membranes, particularly under stress or as a by-product of various redox reactions of metabolic pathways localized in different cellular compartments. However, it is generally agreed that in plant cell chloroplast and peroxisome are most powerful and major source of ROS, particularly under illumination (Apel and Hirt 2004; Bhattacharjee 2005). However, the mitochondria under darkness and in case of nongreen parts appear to be the main source of ROS (Halliwell and Gutteridge 1999). In fact, chloroplast produces both $^1O_2$ and $O_2^{\cdot-}$, H_2O_2, involving PET of the Z-scheme of photosynthesis, whereas mitochondria produce mainly $O_2^{\cdot-}$ at complex I and III of ETC. It is generally estimated that 1–5% of O_2 consumed by isolated mitochondria results in the generation of ROS, depending on the situations (Millar and Leaver 2000).

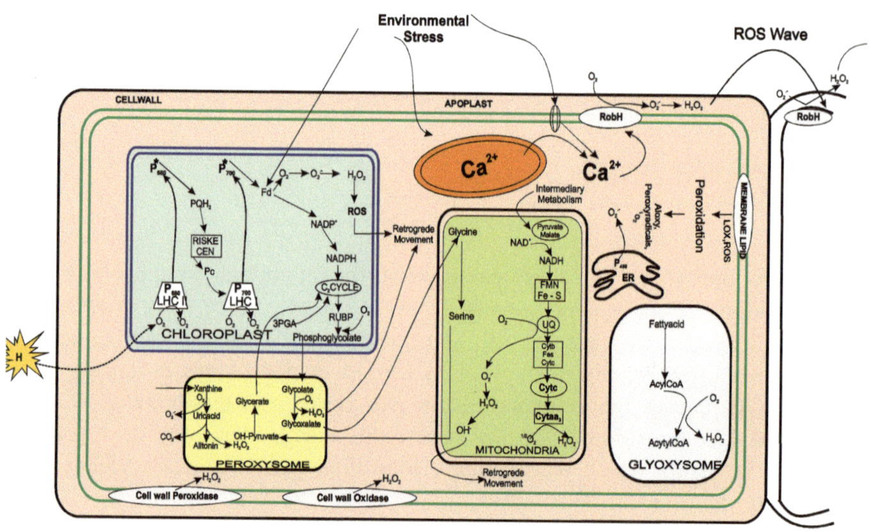

Fig. 1.2 Major subcellular sources and associated formation of ROS wave in plant cell (detail in text)

1.3.1 Chloroplast and Peroxisome-Associated Generation of ROS

The reactive oxygen species arises in plant cells via a number of routes involving most of the cellular compartments (Fig. 1.2). Formation of most of the ROS in plant cells initiates with the formation of superoxide (which arises as a result of single electron-mediated reduction of molecular oxygen principally during the Mehler reactions in chloroplast), followed by its dismutation by SOD to H_2O_2 (Asada 1999). In case of photosynthetic electron transport (PET), spilling electron onto O_2, i.e., photoreduction of O_2 to superoxide (Fig. 1.2), is called Mehler reaction, in honor of the discoverer (Mehlar 1951).

$$2O_2 + 2Fd(red)◆2O2^{\bullet-} + 2Fd(ox)$$

Although photoreduction of oxygen during pseudocyclic electron flow, under excess photochemical energy (EPE), is an important alternative sink for the consumption of excess energy, it is always associated with the risk of generation of toxic ROS (Vanova et al. 2002). If the accumulation of ROS under EPE exceeds the capacity of enzymic and nonenzymic antioxidant systems integrated with PET to remove them, photodynamic damage to photosynthetic apparatus takes place, which leads to cell destruction. The redox imbalance, with the dearth of $NADP^+$ in PSI, causes pseudocyclic electron flow and photoreduction of molecular oxygen triggering the generation of $O_2^{-\cdot}$. The finely tuned and regulated activation of PCRC enzymes and control of rate of Z-scheme electron of photosynthesis are important factors determining the redox state of plant cell. This is extremely important because any imbalance in redox status of the cell causes the donation of electron to O_2, resulting in $O_2^{-\cdot}$. This is primarily because the electron carriers of PSI, particularly ferredoxin, have sufficient negative electrochemical potentials for the reduction of oxygen in absence of $NADP^+$. The majority of $O_2^{-\cdot}$ formed in plant cell in vivo is thought to be produced via electron spilling from reduced ferredoxin to oxygen. Superoxide formed in this way then undergoes dismutation either spontaneously or facilitated by SOD. Superoxide radicals generated are rapidly converted into hydrogen peroxide by chloroplastic Cu/Zn superoxide dismutase. It has been suggested that photoreduction of O_2 to $O_2^{\cdot-}$ and subsequently its scavenging via H_2O_2 to water by ascorbate peroxidase pathway or Halliwell-Asada pathway under EPE may involve about 30% of total electron transport (Bartoli et al. 1999). This also suggests that O_2 plays an important role as an alternative electron acceptor in photoprotection and photooxidative acclimation, particularly under EPE. Therefore, production of large amount of ROS is most likely under EPE, and hence plants evolved efficient strategies by devising antioxidative defense mechanism and integrating them with normal photosynthetic pathway (PET or Z-scheme) to adjust to the imposed oxidative stress (Fig. 1.3).

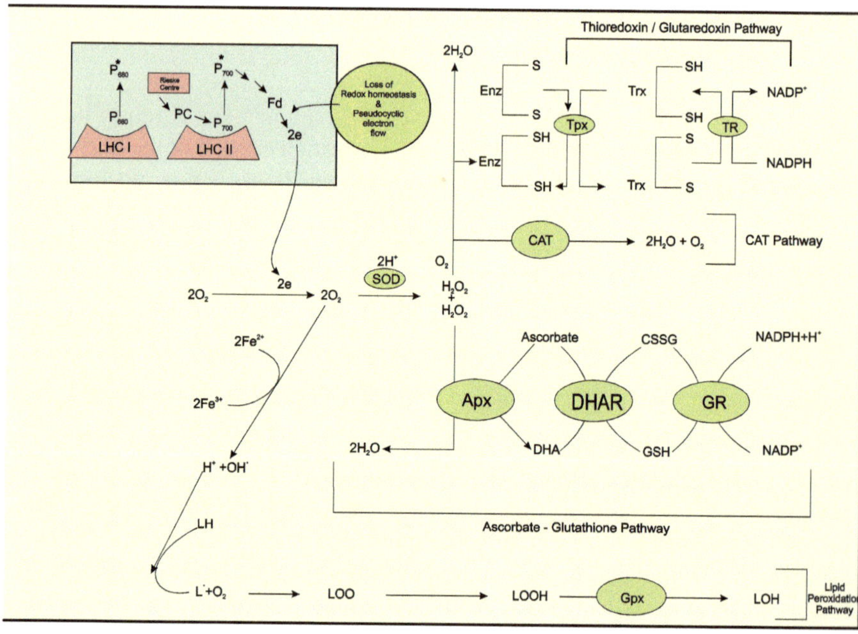

Fig. 1.3 Antioxidative systems integrated with Z-scheme of photosynthesis for regulating the endogenous concentration of ROS in plant cell. Superoxide dismutase (SOD) catalyzes the dismutation of $O_2^{\cdot-}$ to H_2O_2. H_2O_2 is reduced mainly by catalase (CAT) or by ascorbate peroxidase (Apx). Reduction by Apx requires a substance (ascorbate) that is reduced via a cycle or coupled oxidation-reduction reactions catalyzed by monodehydroascorbate reductase (MDHAR) or dehydroascorbate reductase (DHAR) and glutathione reductase (GR), known as Halliwell-Asada pathway or ascorbate-glutathione cycle. Oxidized enzymatic proteins can be reduced by thioredoxin peroxidase (TPx) using thioredoxin [Trx-(SH_2)] as reducing cosubstrate. Oxidized thioredoxin is then regenerated by thioredoxin reductase (TR) at the expense of NADPH+H^+ or reduced ferredoxin. Lipid hydroperoxides (LOOH) that are formed by oxidation of lipids (LH) are reduced by glutathione peroxidase (GPx)

Singlet oxygen, another potent ROS, is incessantly produced during photosynthesis involving mainly PSII. The reaction center complex of PSII consists of heterodimer of D1 and D2 proteins apart from cytochrome b_{559}, enabling the binding of functional prosthetic groups of redox carriers (chlorophyll P_{680}, pheophytin, Q_A, Q_B, etc.). Under EPE the redox state of plastoquinone pool are changed significantly. Q_A and Q_B are found to be over-reduced, causing the oxidized P_{680} to recombine with reduced pheophytin. This condition favors the formation of triplet state of P_{680}, which subsequently transfers the energy to O_2, leading to the generation of singlet oxygen. It is found that excess photochemical energy that leads to photoinhibition of PSII is due to the excessive generation of singlet oxygen (Hideg et al. 1998, 2002).

In C_3 plants, ROS (H_2O_2) may be generated in peroxisome by the oxidation of glycolate through photosynthetic carbon oxidation cycle (PCOC) (Fig. 1.2). In case of PCOC, exhibited by C_3 plants, oxygenation of RuBP by Rubisco constitutes a major alternative sink of electrons, regenerating NADP$^+$, thereby preventing photoinactivation of PSII when CO_2 concentration is reduced or there is acute redox imbalance. Rubisco, in general, favors oxygenation compared to carboxylation as temperature increases, as the equilibrium is shifted more toward oxygen than carbon dioxide. The oxygenation reaction mediated by Rubisco leads to generation of glycolate which on translocation from chloroplast to peroxisomes suffers oxidation, producing the major portion of H_2O_2 as by-product, in photosynthesizing cells (Foyer and Noctor 2003).

Another possible source of production of ROS in plant is chlororespiration. It describes the incomplete reduction of molecular oxygen in presence of respiratory chain consisting of a NADPH dehydrogenase and a terminal oxidase in chloroplast. This system competes with electron transport chain for reducing equivalents. Although this process is associated with algae, more recently the evidence of chlororespiration is also noticed in higher plants, in the form of presence of respiratory chain in chloroplast (Nixon 2000).

Out of all steps, the formation of $O_2^{\bullet-}$ is considered as a rate-limiting step in chloroplast. Once formed $O_2^{\bullet-}$ generates more destructive ROS. It may be protonated to $HO_2^{\bullet}$ on the internal thylakoid lumen or membrane surface or dismutated enzymatically (by SOD) or spontaneously to H_2O_2 on the external "stromal" membrane surface.

1.3.2 Mitochondrial Generation of ROS

ROS like superoxide, hydrogen peroxide, and hydroxyl radicals is also generated actively during mitochondrial electron transport (Fig. 1.2) (Murphy 2009; Vanova et al. 2002; Halliwell and Gutteridge 1999). Formation of $O_2^{\cdot-}$ anions takes place in flavoprotein region of NADH dehydrogenase complex of mitochondrial electron transport (MET). Antimycin A blocks electron flow after ubiquinone in MET and enhance the formation of ROS (Fig. 1.2). In fact, the application of antimycin A causes accumulation of reduced ubiquinone which may undergo auto-oxidation by donating electron to molecular oxygen, resulting in the production of $O_2^{\cdot-}$ (Forman and Boveris 1982). That the mobile electron carrier ubiquinone is the major H_2O_2 generating locations of MET, is supported by several experiments (Boveris and Cadenas 1975; Winston 1990; Maxwell et al. 2002). Apart from the MET, several mitochondrial enzymes can produce ROS. For example, aconitase and 1-galactono-γ lactone dehydrogenase (GAL) are able to feed electrons to MET (Andreyev et al. 2005). $O_2^{\bullet-}$, the primary ROS formed by monovalent reduction of O_2 in the MET, got converted quickly by mitochondrial form of SOD (Mn-SOD) to form relatively stable and membrane-permeable form of ROS, i.e., H_2O_2. H_2O_2 can be further converted to extremely toxic hydroxyl radical (OH•) through Fenton reaction or may be detoxified by APOX. A comparative estimation of ROS level in plant cell (whole leaf point of view) reveals that the peroxisomal and chloroplastic hydrogen peroxide production is the most significant one and may be 30–100 times quicker than the formation of hydrogen peroxide in mitochondria (Fig. 1.4). The relative rates of generation of ROS are shown in Fig. 1.4. Though mitochondrial ROS production is

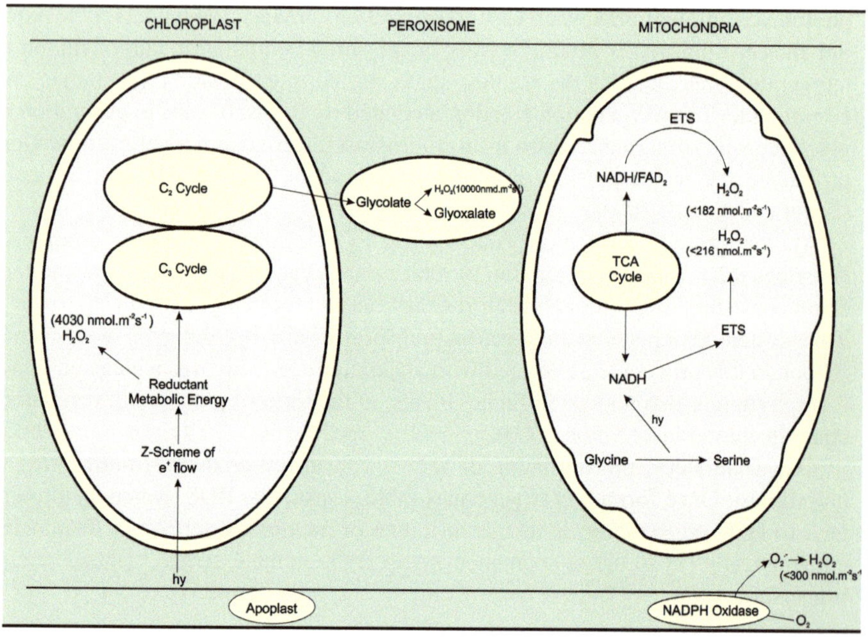

Fig. 1.4 Kinetics of generation of reactive oxygen species H_2O_2 in different compartments of a green cell under ambient environmental condition

not getting influenced in light and dark, but the probability formation of superoxide by MET could be changed on illumination. In fact there are such evidences, when light affects alternate oxidase (Dutilleul et al. 2003). Alternate oxidase is found to influence ROS generation and may determine cell survival under stress (Robson and Vanlerberghe 2002; Maxwell et al. 1999).

1.3.3 Microsomal and Apoplastic Generation of ROS

One of the other source of superoxides in plant cell is NADPH – dependant microsomal electron transport (Gabig 1983). Two potential loci of $O_2^{-\cdot}$ formation in microsomes are auto-oxidation of cytochrome P-450 reductase (a flavoprotein that contains both FAD and FMN) and/or auto-oxidation of oxycytochrome – P-450 complex (Segal and Abo 1993).

Cell wall-associated peroxidase is able to oxidize NADH and in the process catalyze the formation of $O_2^{\cdot-}$. This enzyme makes use of H_2O_2 to catalyze the oxidation of NADH to NAD^+, which subsequently reduces O_2 to $O_2^{\cdot-}$ (Bolwell et al. 1995). Superoxide consequently might undergo dismutation by SOD to produce H_2O_2 and O_2.

Cell membrane-localized NADPH oxidase was identified as a major contributor to their bactericidal capacity (Segal and Abo 1993). Apart from NADPH oxidase, germin-like oxalate oxidases, pH-dependent cell wall peroxidases, and amine oxidases are also potential sources of ROS H_2O_2 in apoplast of plant cell. It is found that the pH-dependent cell wall peroxidases are specifically activated in alkaline pH and in presence of a suitable reductant produces H_2O_2. Therefore, it is the alkalization of apoplast which took place upon elicitor recognition that precedes the oxidative burst and the production of ROS H_2O_2 by activating the pH-dependent cell wall peroxidases during biotic stress (Bolwell et al. 1995).

1.3.4 NADPH Oxidase (RBOH)-Induced Formation of ROS

The plasma membrane-bound NADPH oxidases, otherwise called as respiratory burst oxidase homologs (RBOHs), have been proposed to play a fundamental role in the oxidative burst and also production and accumulation of ROS under stress conditions as well as during developmental processes in plants (Pei et al. 2000; Torres et al. 2002; Vanova et al. 2002; Kwak et al. 2003; Apel and Hirt 2004; Laloi et al. 2004). The apoplastic enzyme transfers electrons from cytoplasmic NADPH to O_2 to form ROS $O_2^{\cdot-}$, which subsequently gets dismutated to H_2O_2. NADPH oxidase is primarily involved in plant-biotic and abiotic stress response (Sagi and Fluhr 2001; Torres et al. 2002). The role of NADPH oxidase can be vouched from the study with application of diphenyleneiodonium (DPI, an important inhibitor of NADPH oxidase) which impairs H_2O_2 production during stress conditions in plants (Overmyer et al. 2003; Laloi et al. 2004). It is found that the RBOH-dependent $O_2^{\cdot-}$ production is associated with the membrane lipid peroxidation and apoptosis (Neill et al. 2002). NADPH oxidases instigate cell damage by initiating ROS production and evoke oxidative stress and associated reduction in growth under extreme environmental stresses. The role of NADPH oxidase has been shown to be further elaborated, particularly during oxidative burst, when this enzyme initiates Ca^{2+} signaling and the activity of MAPK in downstream signaling cascades. The enzyme is also found to influence significantly the signaling pathways of growth regulators, such as ethylene (ET), salicylic acid (SA), and jasmonates (JA), through redox signaling and may subsequently regulate the defense physiology of plants (Overmyer et al. 2003; Evans et al. 2005).

Structurally, the NADPH oxidases of plant origin have cytosolic C-terminal domain having FAD and NADPH-binding sites, followed by six conserved domain with trans-membrane-spanning peptide and a cytosolic N-terminal extension domain having a phosphorylation target site and two Ca^{2+}-binding motifs (EF-hand) (Kimura et al. 2012; Drerup et al. 2013). Once this enzyme gets activated (under stress or developmental event), O_2 is reduced to superoxide ($O_2^{\cdot-}$) in the apoplast which eventually gets further dismutated to H_2O_2, spontaneously or catalytically by SOD (Lin et al. 2009).

1.3.5 Cell Wall and Other Enzymatic Sources of ROS

Plant cell walls play an important role in defense mechanisms against pathogens and in the degradation or compartmentalization of some xenobiotic chemicals.

Though, most of the reactions in these sites are biosynthetic in nature, being a dynamic site of metabolism, in several instances the cell wall exhibit O_2 utilization and activation. Most significantly the cell wall exhibit H_2O_2-dependent reactions, which randomly link the phenylpropanoid subunits to form secondary wall material lignin (Higuchi 2006). Similarly, it is also the site of redox reactions involving the enzyme malate dehydrogenase and NADH oxidase. In first redox reaction, NADH may be formed by cell wall-associated enzyme malate dehydrogenase which subsequently gives rise to ROS H_2O_2 by membrane-associated NADH oxidase. Another cell wall-associated enzyme diamine oxidases is also found to involve in the production of ROS in the apoplast. Polyamines can also generate peroxides in apoplast through the auto-oxidation of quinone (Spiteller 2003; Higuchi 2006).

So far as enzymatic source is concerned, glycolate metabolism involving glycolate oxidase during PCOC is the most active source of generation of ROS H_2O_2. ROS can also be formed by other metabolic events involving oxidases. For instance, the oxidative degradation of purines by xanthine oxidases, polyamines by polyamine oxidases, and polyphenols by polyphenol oxidases are the most important examples of enzymatic formation of ROS. In all cases H_2O_2 is produced as by-product (Halliwell and Gutteridge 1999). Enzymatic sugar oxidation can also generate ROS, as noticed in some fungi (Apel and Hirt 2004).

1.4 Membrane Lipid Peroxidation: A Prospective Source of Oxyfree Radicals in Plant Cell

One of the major example and primary source of carbon-centered ROS production in plant cells is the peroxidation of lipids. The major substrate used for this event is primarily the phospholipids and other forms of lipids associated with the cell membranes (Hameed et al. 2013; Winston 1990; Bhattacharjee 2014). Membrane lipid peroxidation (MLP) may be enzymatic and nonenzymatic. In case of enzymatic MLP, lipoxygenase mediates process in plant cells. The process can also be initiated by a number of ROS itself. From the mechanistic point of view, the entire process of MLP takes place through three separate stages (Fig. 1.5): initiation, progression, and termination. MLP can generate a variety of products depending on the nature of substrates (fatty acids) and oxidants used in the process as well as the severity of the oxidation event (Sharma et al. 2012; Hameed et al. 2013; Bhattacharjee 2014). However, in most of the cases, MLP produces the products having moieties containing aldehydes, ketones, hydroxyls, hydroperoxyls, carboxylic acids, and the carbon skeletons with trans-double bonds (Borchman and Sinha 2002).

At the outset, membrane lipids need to be hydrolyzed involving specific lipases, like phospholipase C. The activity of these classes of phospholipases enhanced significantly under diverse types of environmental stresses. Hydrolysis of membrane lipid, in general, causes formation of free fatty acid. But, out of these, only those fatty acids having cis,cis-1,4-pentadiene moiety in their skeletons became the substrate for the enzyme lipoxygenase. The activated lipoxygenase commences the initiation event involving the transition metal complexes (Fe and Cu), causing the

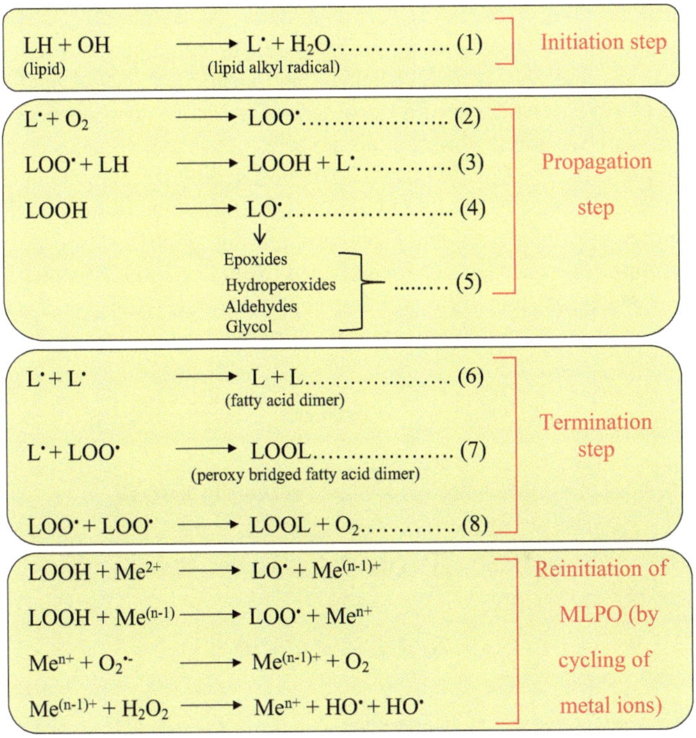

Fig. 1.5 Chemical events associated with membrane lipid peroxidation (MLPO): a potential basis of ROS formation in plant cell

generation of lipid alkyl radicals. The role of these transition metal complexes in the initiation process is extremely significant either as a catalyst in the degradation of existing lipid hydroperoxides or can initiate in the formation of activated oxygen complex capable of abstracting allylic hydrogens from free fatty acids having cis,cis-1,4-pentadiene moiety.

In nonenzymatic MLP, OH· is the most potent ROS capable of initiating the process. Other forms of ROS, like superoxide, hydrogen peroxide, etc., can also initiate MLP though not as efficiently as OH·. Most significantly, OH· is found to initiate MLP of mitochondrial PUFAs by abstracting a hydrogen atom with the generation of lipid aldehydes, alkenals, and hydroxyalkenals (HNE and MDA), most of which are found to be extremely cytotoxic and capable of initiating apoptosis (Rhoads et al. 2006; Bhattacharjee 2014). Loosely associated transition metal complexes (Fe and Cu) can also catalyze the degradation of lipid peroxides, causing the formation of secondary ROS, like alkoxy and peroxy radicals, which can further initiate the fresh round of chain reactions of lipid peroxidations (Winston 1990; Aust et al. 1985, Fig. 1.5).

Free fatty acids, particularly the PUFAs with cis,cis-1,4-pentadiene moiety, are more susceptible to oxidative MLPO by ROS because of the abundance of double bonds in a fatty acid skeleton and the easy oxidation process through elimination of a hydrogen atom (Porter et al. 1995). As lipoperoxides are extremely unstable, they are catabolized to form a number of secondary products including the reactive carbonyl compounds, particularly aldehydes [such as malondialdehyde (MDA) and 4-hydroxy-2-nonenal (4-HNE)], which in turn cause serious oxidative consequences to cells by interacting with free amino groups of amino acids of proteins (Sochor et al. 2012).

This entire process of MLPO can also be initiated by the enzyme lipoxygenase, instead of ROS (LOX, Fig. 1.6). LOX initiates the formation of fatty acid hydroperoxides and swings the peroxidation process. During aging, senescence, and stress responses, lipoxygenases (LOX) are activated (Spiteller 2003; Bhattacharjee 2005, 2014). Lipoxygenases transform PUFAs (with $-CH = CH - CH_2 - CH = CH -$ moiety) in a reaction called lipid peroxidation to lipid hydroperoxides (LOOHs), which

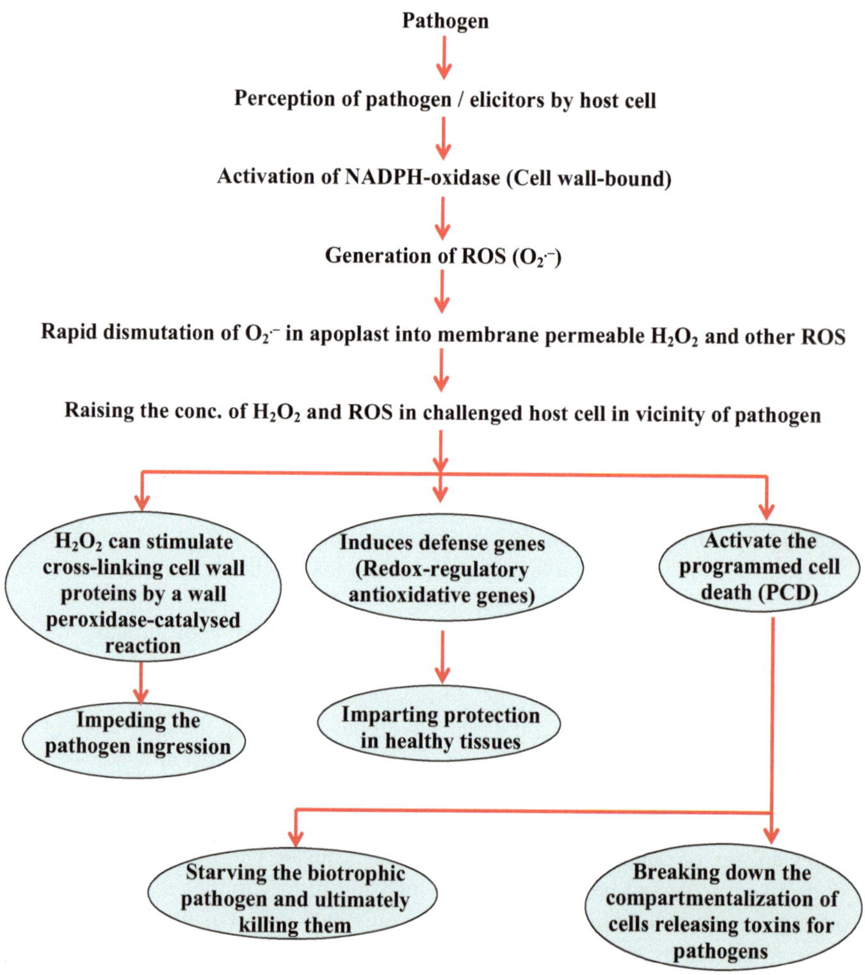

Fig. 1.6 Hypothetical integrative model for function of ROS generated during plant-pathogen interaction

being extremely unstable further decomposed to a great variety of secondary products. In fact, LOX is capable of removing in a regio- and stereospecific manner a hydrogen atom from a double allylically activated methylene carbon of PUFA. In subsequent event, the hydrogen atom, while remaining attached with the enzyme, reacts with the complex-bound Fe^{3+} in the active center of LOX through the formation of a H^+ and a Fe^{2+} ion. The lipid radical L· adds oxygen and generates a peroxyl radical (LOO·). Afterward Fe^{2+} donates electrons to the peroxyl radical and produces a peroxyl anion (LOO^-). LOO^- then combine with the proton to form LOOH (de Groot et al. 1975, Fig. 1.5). However, the most important point in this aspect to note is that during enzyme-catalyzed MLPO, the peroxyl radical always remains attached with the enzyme complex and unable to escape.

There seems to be a strong connection between MLPO and senescence, as it is evident that the MLPO products increase with age approaching (Jabs 1999). Similarly, secondary MLPO products are detectable under biotic and mechanical stress and wounding of plant tissue (Spreitzer et al. 1989). MLPO is also associated with the formation of the plant signaling compounds like jasmonic acid and eicosanoids in mammals. There seems to be lot of similarity between MLPO chemistry and physiology of plants and mammals. Aging in mammals, like plants, is characterized by dramatic increase in MLPO products and other encoded proteins which induce LOX activity. Another way to monitor LPO was by assessing the change in composition of PUFAs and the alteration of the ratio between saturated fatty acids and PUFAs. As the quantitative amount of saturated fatty acids remains unaffected in LPO processes, any change in PUFAs might hint at the LPO, as observed during growth and senescence processes.

1.5 Abiotic Stress and Production of ROS

Fluctuating environmental condition such as extremes of temperature, UV radiation, soil water availability, intensity of light, presence of salts, excess ions, heavy metal exposure, and air pollution can lead to loss of redox homeostasis due to increased production and accumulation of prooxidants and hence lead to unwanted oxidative damage in plant cells (Alscher et al. 1997; Eltsner 1987; Bartoli et al. 1999, Bhattacharjee and Mukherjee 1996, 2001, 2002, 2003; Bhattacharjee 2005, 2008; Apel and Hirt 2004; Imlay 2008; Mittler 2017). Water scarcity due to drought and salinity upregulates the formation of ROS in a number of ways. Underperformance of PCRC, coupled with the adverse changes in photosystem activities or photosynthetic electron transport capacity under such conditions, results in pseudocyclic electron flow and accelerated generation of ROS via the Mehler reaction (Asada 1999; Sharma et al. 2013). Under drought or salinity, CO_2 assimilation is limited due to significantly reduced stomatal conductance (primarily required for the maintenance of water use economy), causing dearth of $NADP^+$. Due to the deficiency of electron acceptor $NADP^+$ and the over- energization of the photosynthetic Z-scheme, significant leakage of electrons to O_2 by the Mehler reaction takes place. According to Biehler and Fock (1996), a significant amount of photosynthetic electrons, as much as 50%, is spilled to the Mehler reaction under water deficit condition in wheat plants, compared to unstressed plants. Further, for the dissipation of surplus absorbed photochemical energy in the PSII core, the overexcited antenna system interacts with molecular O_2, leading to the generation of $^1O_2$, a highly detrimental ROS produced under such conditions (Foyer and Harbinson 1994). It is also observed that under dehydration and temperature stress, the photorespiratory or C_2 pathway enhanced, especially when there is dearth of CO_2 due to reduced stomatal conductance, reducing partial pressure of CO_2 more as compared to O_2. It is surprising but fact that a single metabolism i.e. photorespiration is likely to contribute over 70% of total H_2O_2 production under dehydration stress (Noctor et al. 2002). In this entire event, abscisic acid plays the central role, as it downregulates the turgor pressure of guard cells for the reduction of water loss, subsequently reducing the availability of CO_2 and photosynthetic carbon reduction cycle (PCRC). The reduction in the rate of PCRC, on the other hand, reduces the regeneration of $NADP^+$, which leads to the initiation of pseudocyclic electron flow and consequently comes up with the formation of ROS from PSII and PSI. This kind of abiotic stress-induced ABA-mediated loss of redox homeostasis of the chloroplastic cell due to overaccumulation of ROS and subsequent enhancement of photooxidative damage is a universal feature of abiotic stress-induced metabolic dysfunction, cytotoxicities, and injuries (Jiang and Zhang 2001; Bhattacharjee 2005, 2014).

Salinity, in general, imposes oxidative stress to plant tissue by restricting electron mobilization of chloroplasts and mitochondria, causing redox imbalance, necessary for the formation of ROS. Salt stress can also cause an induction of PCOC or photorespiration (Hernández et al. 2000). The basic mechanism of upregulation of PCOC and associated formation of ROS is quite similar to that of drought stress, i.e., reduction of ABA-induced stomatal conductance and availability of CO_2 and shifting the metabolism from PCRC to PCOC. Salinity also causes shifting of photosynthetic electron transport system from noncyclic and cyclic type to pseudocyclic type, due to the dearth of $NADP^+$, leading to enhanced formation of ROS and oxidative damage.

Elevated concentrations of heavy metals can seriously perturb the redox homeostasis of the plant cell and impose oxidative damage by disturbing photosynthetic electron flow (Somashekaraiah et al. 1992; Van Assche and Clijsters 1990). For example, Cd^{2+} inhibits PSII-related electron transport and causes redox imbalance. Further its reaction with –SH group of different ETC-related protein and enzyme aggravates further the synthesis of ROS and metabolic dysfunction (Halliwell and Gutteridge 1984). Hyper-accumulation of heavy metals was shown to be involved in the induction of secondary oxidative stress in a heavy metal-sensitive *Phaseolus*, *Holcus lanatus*, and *Amaranthus* (Somashekaraiah et al. 1992; Hendry et al. 1992; Bhattacharjee 1998). In fact, the accumulation of borderline heavy metals cause significant upregulation in the synthesis of ROS probably by inhibition electron flow through photosystem II. *Amaranthus lividus* seedlings exposed to different magnitude of heavy metal stress (Pb^{2+} and Cd^{2+}) cause significant oxidative membrane damage by stimulating ethylene formation, lipid peroxidation, and metabolic dysfunction (Bhattacharjee and Mukherjee 1996; Bhattacharjee 1998). Similarly, Fe toxicity induces oxidative stress in plants by upregulating the formation of most potent ROS, OH from $O_2^{\cdot-}$ and H_2O_2. The transition metal ions like Cu^{2+} and Fe^{2+}, if present in excess, catalyze Fenton-type reactions for the formation of OH·. Some workers observed an upregulation of enzymatic MLPO and downregulation of competence of antioxidative defense mechanism as the probable mechanism of heavy metal-induced oxidative stress in plants (Somashekaraiah et al. 1992; Bhattacharjee, 1998).

Herbicide exposure to plant can also induce secondary oxidative stress either by direct involvement in ROS formation or metabolic dysfunction. Paraquat exposure induces inhibition of Z-scheme of electron flow and causes formation of ROS (Bruke et al. 1985). Herbicides inhibiting the normal flow of electrons during operational Z-scheme, often exhibit PSII-mediated reduction of plastoquinone, which subsequently forms a monocation radical, capable to react with molecular oxygen to produce $O_2^{\cdot-}$ and other ROS (Eltsner 1982; Arora et al. 2002).

Extremes of temperature, both hyperthermia and hypothermia, play significant role in changing the redox homeostasis of the plant cell. Both imbibitional heat and chilling stress to a tropical leaf crop *Amaranthus lividus* impose secondary oxidative stress to the germinating tissue by upregulating the formation of ROS and downregulating antioxidative defense mechanisms (Bhattacharjee 2005, 2008; Bhattacharjee and Mukherjee 2003). Chilling stress leads to loss of redox homeostasis by exacerbating imbalance between light absorption and light use by inhibiting PCRC, enhancing photosynthetic ROS formation (Fadzillah et al. 1996). Extremes of temperature also downregulate the transcriptional activation of *rbcL* and *rbcS* genes, causing significant reduction in RUBISCO pool and activity, eventually leading to significantly higher electron flux to O_2 associated with the formation of ROS (Zhou et al. 2006). As an evidence of origin of oxidative stress, ROS accumulation in chloroplast is found to be negatively correlated with the early RUBISCO kinetics and the rate of PCRC (Prasad et al. 1994).

Another significant cause of oxidative stress to plant due to elevated production of ROS is the exposure to UV-B radiation. It is found that excess exposure of UV-B to crops considerably diminish net photosynthetic rate. UV-B, in general, causes significant downregulation in light-saturated rate of PCRC, along with reduction in kinetics of RUBISCO carboxylation (Allen et al. 1997). The impact on photosynthesis by UV-B can be vouched by a significant reduction in the proportions of maximum chlorophyll fluorescence yield and quantum yield of photosynthetic oxygen formation in leaves of rice. This reduction in PCRC due to UV-B is the sole cause of formation of ROS responsible for oxidative damage in plants (Strid et al. 1994). Rao et al. (1996) and Han et al. (2009) proposed that UV-B exposure upregulates the activities of NADPH oxidase and downregulates activities of antioxidative defense enzyme, causing the formation of ROS and loss of redox of the cell.

1.5.1 Generation of ROS Under Biotic Stress

Biotic stress or infection to plants significantly alters the redox status by aggravating the formation of ROS (Fig. 1.6). ROS are produced by plant cells when they "sense" the presence of infectious pathogen trying to invade the epidermal tissue (Levine et al. 1994). It is one of the fastest plant responses to pathogens. For instance, ROS burst is detected within a few minutes following the addition of pathogen elicitor (Legendre et al. 1993). Similarly, the contact of cell wall-degrading enzymes (mainly of pathogen origin) elicits a rapid ROS burst in host cells (Brady and Fry 1997). The oxidative burst in plant cells exposed to pathogen-based elicitor is a generalized fact which may even take place in response to herbivore attack (Bi and Felton 1995).

Bacterial pathogens, in general, exhibit two bursts of ROS in a temporal manner:

(i) An early transitory oxidative burst (within 30 min of inoculation that may persist for another 30 min)
(ii) A delayed but more pronounced oxidative burst (approximately 4–6 h post inoculation which persists for few hours) (Levine et al. 1996)

The first oxidative burst constitutes a general plant response to an assault caused by elicitor. It is a spontaneous defense response irrespective of nature of invading organism, i.e., virulent, avirulent, or saprophytic. However, the second oxidative burst is absolutely associated with incompatible interactions with avirulent bacteria. And it is the second burst of ROS, which play a pivotal role in holding back the bacterial growth by inducing HR (hypersensitive response mechanism) (Levine et al. 1994).

The multiple functions of oxidative burst associated with plant-pathogen interaction have been presented in Fig. 1.6. Within a few minutes of infection, H_2O_2 facilitates required cross-linking of cell wall proteins by a peroxidase-catalyzed reaction, impeding further pathogen entrance (Levine et al. 1994). Further enhancing titer of ROS H_2O_2 induces a signaling cascade upregulating a subset of inducible defense genes (Bhattacharjee 2005; Mahalingam and Fredroff 2003). Once the endogenous titer of ROS reaches a certain threshold value, cell wall initiates programmed cell death (PCD), which in turn exhibited the following important roles during pathogenesis:

(a) PCD destroys the surrounding cell of the ingressed pathogen, depriving them from having essential nutrients, causing starvation of biotrophic pathogens (Lamb and Dixon 1997).
(b) After the breakdown of compartmentalization of host cell, toxic antimicrobial chemicals and ROS spilled out of the vacuole and cells further poisoning the pathogens. The high local concentrations of ROS, persistent for several hours, further destruct the microbial pathogen (Devlin and Gustine 1992).

The NADPH oxidase or RBOH similar to that of mammalian system seems to be related with the generation of ROS and associated PCD in plants. NADPH oxidase catalyzes the formation of ROS superoxide by one electron-mediated reduction of O_2 using NADPH as source of electron donor

$$2O_2 + 2NADPH + H^+ \xrightarrow{\text{NADPH oxidase}} O_2^- + 2NADP^+ + 2H^+$$

This starting event of superoxide formation subsequently became the source of other ROS. Knockout mutations of two *Arabidopsis* rboh genes (rboh D and Atrboh F) significantly reduce ROS formation associated with the disease resistance of *Arabidopsis* to avirulent pathogens. This experiment provides the evidence for the direct role of plant NADPH oxidase associated with disease resistance (Torres et al. 2002; Doke et al. 1994).

As an alternative mechanism, several peroxidases residing in apoplastic space may contribute in the formation of ROS. These apoplastic peroxidases evoke their action under different circumstances. In the first instance these peroxidases, in presence of ROS and phenolic substances, induce a peroxidative cycle for the synthesis of lignin and other secondary polyphenolic compounds of cell wall polymers. On the other hand, if they are substituted by NADPH or other redox component or reductants, reaction provides the molecular prerequisite for the formation of H_2O_2 generating NADPH oxidase or RBOH activity (Chen and Schopfer 1999; Ogawa et al. 1997)

1.6 ROS Wave

The sustained and self-propagating nature of ROS formation, mainly mediated by the activation of *Respiratory Burst Oxidase Homolog D* (RBOHD) in each cell along its systemic path through apoplast, is being termed as ROS wave (Miller et al. 2009; Mittler 2017) (Fig. 1.2). Discovered initially by Miller et al. (2009) and subsequently extensively reviewed by Miller et al. (2010), Gilroy et al. (2014), and Mittler and Blumwald (2015), it is found that ROS wave is one of the main causes of systemic stress signal. It is estimated that this ROS wave can migrate at the rate of 8.4 cm/min and may be straight way linked to the calcium wave (Gilroy et al. 2014) and possibly to other forms of electric signals (Suzuki et al. 2013; Fig. 1.2). The systemic stress acclimation largely depends on the ROS wave as it does not communicate specificity to the systemic response of plants to different environmental assaults (Suzuki et al. 2013; Miller et al. 2010). This wave is found to be integrated with other metabolic/signaling pathways to enable plant rapid systemic acclimation response for growth and development (Suzuki et al. 2013; Mittler and Blumwald 2015). For example, the initiation of ROS wave by hyperthermia was shown to improve the heat stress acclimation of systemic tissues, which is largely found to be regulated by spatiotemporal interaction or cross talking with ABA signaling (Mittler and Blumwald 2015). Application of excess photochemical energy to certain leaf tissue, resulted in initiation of a ROS wave that enabled systemic tissues in far reaching places to withstand further episodes of excess photochemical energy or light stress. This phenomenon of

ROS wave induced systemic signaling was found to be accompanied by metabolic adjustment, like the accumulation of photorespiratory glycine and serine, in non-stressed systemic tissues (Suzuki et al. 2013)

1.7 ROS Generation During Senescence

Several evidence directly implicate ROS in the natural course of senescence (Dhindsa et al. 1982; Thompson et al. 1987; Arora et al. 2002). Lipid peroxidation (LPO), primarily membrane lipid peroxidation (MLPO), plays a significant role during the process (Thompson et al. 1987; Arora et al. 2002). MLPO being an active source of ROS, like alkoxy, peroxy radicals, singlet oxygen, etc., aggravates further deterioration of cell, when initiated. In fact, MLPO associated with senescence is initiated by ROS or lipoxygenase. There are evidences of significant upregulation in LOX activity with advancing senescence (Thompson et al. 1987). LOX play a central role in promoting oxidative disassembly of free fatty acids of lipids during senescence, as it can initiate chain reaction of lipid peroxidation and can also form $^1O_2$. During senescence the LOX-mediated MLOP increases in a temporal pattern consistent with its putative role in promoting oxidative injury (Thompson et al. 1987; Grossman and Leshem 1978). Mobilizing free fatty acids from membrane phospholipids, some of which serves as substrate lipoxygenase, is also noticed. As an evidence, a significant reduction in membrane phospholipid has been manifested as an increased sterol: fatty acid ratio, during early stages of senescence, has been noticed (Thompson et al. 1987). Membrane-bound lipases play a critical role in mobilizing free fatty acids from membrane lipids during senescence. The lipid-degrading enzymes, like phospholipase D, phosphatidic acid phosphatase, and lipolytic acyl hydrolase, are found to be directly with the senescence of microsomal membrane (Thompson et al. 1987). Another evidence of ROS involvement during senescence of plant cell is the enhancement in the pool of Fe that could be used to catalyze the formation of ROS during the event. Though it is not known convincingly whether Fe pool during senescence increases due to decompartmentalization or proteolytic degradation of metalloproteins, its involvement in the formation of ROS associated with senescence is well established.

1.8 Conclusion and Perspective

The underlying principle of this chapter describing physiological basis of ROS formation in relation to its subcellular topography in plant has its root in large number of physicochemical aberrations or conditions in which ROS has been implicated with their variety of response. In fact, ROS are constant companion of aerobic metabolism and incessantly produced as by-products of redox chemical reactions (membrane-linked electron transport of chloroplast, mitochondria, plasma membrane, etc.) in different cellular compartments. Under normal environmental condition, plant maintains redox homeostasis, as the ROS production in various cellular compartments is low and being continuously scavenged by their counterpart of the elaborate redox network, i.e., antioxidative defense system confined to different cellular compartments. However, the redox homeostasis or equilibrium between the production of ROS and their subsequent scavenging often found to be perturbed under fluctuating environmental conditions, causing imposition of secondary oxidative stress and associated damages to plant cell. It is also noticed that plants are also capable of generating ROS purposefully (oxidative burst) by activating various oxidases/peroxidases (RBOH) to combat environmental stresses, both abiotic and biotic. However, plant responds to various unfavorable environmental cues by modulating ROS-antioxidant interaction at metabolic interface and subsequently forming internal redox cue. Not only that, depending on the nature, severity, and duration of environmental stresses, plants differentially modulate redox cue by tightly regulating the endogenous titer of ROS. In fact, the evil consequence of oxidative stress still overshadows much of our current viewpoint and understanding of ROS generation and functions. While many questions remain to be addressed regarding their generation, like subcellular regulation of titer, communication and corresponding control ROS signaling in plant cells, etc., it is clear that ROS production has vital relevance to physiological and developmental status of plant cells. However, if one accepts the notion that differences in regulation of spatiotemporal synthesis of ROS through its different chemotypes determine the type of response, it is extremely tough to explain how and what delineate between the lethal and adaptive responses of ROS.

Though significant progress has been made in ROS Biology in recent years, there are many lacunae in our knowledge regarding the formation and effect of ROS on plants, mainly due to their short half-life and high reactivity. Advanced cutting-edge analytical techniques will definitely assist us in the future to unearth the unexplored dynamic mechanism of formation and fate of ROS that will help us in developing broader view of the role of ROS in plants. Integration of the concept of genomics, metabolomics, and proteomics with ROS Biology will help in clear understanding of complicated redox network involved in cellular responses to environmental stress and developmental processes.

References

Allen DJ, McKee IF, Farage PK, Baker NR (1997) Analysis of the limitation to CO_2 assimilation on exposure of leaves of two *Brassica napus* cultivars to UV-B. Plant Cell Environ 20:633–640

Alscher RG, Hess JL (1993) Antioxidant in higher plants. CRC Press, Boca Raton. ISBN 0-8493-6328-4

Alscher RG, Donahue JL, Cramer CL (1997) Reactive oxygen species and antioxidants: relationship in green cells. Physiol Plant 100:224–233

Andreyev AY, Kushnareva YE, Starkov AA (2005) Mitochondrial metabolism of reactive oxygen species. Biochemistry 70:200–214

Apel K, Hirt H (2004) Reactive oxygen species; metabolism, oxidative stress and signal transduction. Ann Rev Plant Biol 55:373–399

Arora A, Sairam RK, Srivastava GC (2002) Oxidative stress and antioxidative system in plants. Curr Sci 82(10):1227–1273

Asada K (1999) The water-water cycle in chloroplast:scavenging oxygens and dissipation of excess protons. Annu Rev Plant Physiol Plant Mol Biol 50:601–639

Asada K, Takahashi M (1987) Production and scavenging of active oxygen in photosynthesis. In: Kyle DJ, Osmund CB, Arntzen CJ (eds) Photoinhibition. Elsevier, Amsterdam, pp 227–287

Aust SD, Moorehouse CE, Thomas CE (1985) Role of metals in oxygen radical reactions. J Free Radic Biol Med 01:03–25

Bartoli CG, Simontacchi M, Eduardo T, Beltrano J, Montaldi E, Puntarulo S (1999) Drought and watering-dependent oxidative stress: effect on antioxidant content in *Triticum aestivum* L. leaves. J Exp Bot 50(332):375–383

Bhattacharjee S (1998) Membrane lipid peroxidation, free radical scavengers and ethylene evolution in *Amaranthus* as affected by lead and cadmium. Biol Plant 40(1):131–135

Bhattacharjee S (2005) Reactive oxygen species and oxidative burst: role in stress, senescence and signal transduction in plants. Curr Sci 89(05):1115–1121

Bhattacharjee S (2008) Calcium –dependent signaling pathway in the heat induced oxidative injury in Amaranthus lividus L. Biol Plant 52(01):137–140

Bhattacharjee S (2012) An inductive pulse of hydrogen peroxide pretreatment restores redox-homeostasis and mitigates oxidative membrane damage under extremes of temperature in two rice cultivars (*Oryza sativa* L., cultivars Ratna and SR 26B). Plant Growth Regul 68:395–410

Bhattacharjee S (2014) Membrane lipid peroxidation and its conflict of interest: the two faces of oxidative stress. Curr Sci 107:1811–1823

Bhattacharjee S, Mukherjee AK (1996) Lead and cadmium mediated membrane damage in Rice. II. Hydrogen peroxide level, superoxide-dismutase, catalase and peroxidase activities. J Ecotoxicol Environ Monit 06(01):035–039

Bhattacharjee S, Mukherjee AK (2001) Abiotic stress induced membrane damage in plants: a free radical phenomenon. In: Pandey SK (ed) Advances of stress physiology of plants. Scientific Publishers of India, Jhodpur, pp 16–36

Bhattacharjee S, Mukherjee AK (2002) Salt stress induced cytosolute accumulation, antioxidant response and membrane deterioration in three rice cultivars during germination. Seed Sci Technol 30:279–287

Bhattacharjee S, Mukherjee AK (2003) Implication of reactive oxygen species in heat shock induced germination and early growth impairment in *Amaranthus lividus* L. Biol Plant 47(04):517–522

Bhattacharjee S, Mukherjee AK (2004) Heavy metal induced germination and early growth impairment in *Amaranthus lividus* L.: implications of oxidative membrane damage. J Plant Biol 31(1):01–11

Bi JL, Felton GW (1995) Foliar oxidative stress and insect herbivory: primary compounds secondary metabolites and ROS as components of induced resistance. J Chem Ecol 21:1511–1530

Biehler K, Fock H (1996) Evidence for the contribution of the Mehler-peroxidase reaction in dissipating excess electrons in drought-stressed wheat. Plant Physiol 112:265–272

Bolwell GP, Bnti VS, Davis DR, Zinmorlin A (1995) The origin of oxidative burst in plants. Free Radic Res 23:517–532

Borchman D, Sinha S (2002) Determination of products of lipid oxidation by infrared spectroscopy. In: Armstrong D (ed) Oxidative stress biomarkers and antioxidant protocols. Humana Press Inc, Totowa, pp 21–28

Boveris A, Cadenas E (1975) Mitochondrial production of superoxide anions and its relationship to the antimycin insensitive respiration. FEBS Lett 54(3):311–314

Bowler C, Van Camp W, Van Montagu M, Inzé D (1994) Superoxide dismutases in plants. Crit Rev Plant Sci 13:199–218

Brady JD, Fry SC (1997) Formation of di–isodityrosine and loss of isodityrosine in cell walls of tomato of cell suspension cultures treated with fungal elicitors or hydrogen peroxide. Plant Physiol 115:87–92

Bruke JJ, Gamble PE, Hatfield JL, Quinsberry JE (1985) Morphological and biochemical responses to field water deficit. I. Responses to glutathione reductase activity and paraquat sensitivity. Plant Physiol 79:415–419

Buchanon BB, Balmer Y (2005) Redox regulation: a broadening horizon. Annu Rev Plant Biol 56:187–220

Chen SX, Schopfer P (1999) Hydroxyl radical production in physiological reactions: a novel function of peroxidase. Eur J Biochem 260:726–735

Chowdhury FK, Rivero FM, Blumwald E, Mittler R (2016) Reactive oxygen species, abiotic stress and stress combination. Plant J. https://doi.org/10.1111/tpj.13299

Cressien G, Firmin J, Fryer M, Kular B, Leyland N (1999) Elevated glutathione biosynthesis capacity in the chloroplast of transgenic tobacco plants paradoxically causes increased oxidative stress. Plant Cell 11:1277–1292

Das K, Roychoudhury A (2014) Reactive oxygen species (ROS) and response of antioxidants as ROS-scavengers during environmental stress in plants. Front Environ Sci 2:53

Dat J, Van Breusegerm F, Vandenabeele S, Vranova E, Van Montagu M, Inze D (2000) Active oxygen species and catalase during plant stress response. Cell Mol Life Sci 57:779–786

Davison PA, Hunter CN, Horton P (2002) Overexpression of β-carotene hydroxylase enhances stress tolerance in Arabidopsis. Nature 418:203–206

De Groot JJMC, Veldink GA, Vliegenthart JFG, Bpldingh J, Wever R, Vangelder BF (1975) Demonstration by EPR spectroscopy of functional role of iron in soybean lipoxygenase 1. Biochem Biophys Acta 377:71–79

Del Rio LA, Sandalio LM, Palma JM, Bueno P, Cospus FJ (1992) Metabolism of oxygen radicals in peroxisomes and cellular implications. Free Radic Biol Med 13:557–580

Devlin WS, Gustine DL (1992) Involvement of oxidative burst in phytoallexin accumulation and hypersensitive reaction. Plant Physiol 100:1189–1195

Dhindsa RJ, Dhindsa PP, Thorpe TA (1982) Leaf senescence and lipid peroxidation: effect of some phytohormones and scavengers of free radicals and singlet oxygen. Physiol Plant 56:543–557

Doke N, Miura Y, Leandro MS, Kawakitn K (1994) Involvement of superoxide in signal transduction: responses to attack pathogens, physical and chemical shocks and UV irradiation. In: Foyer CH, Mullineaux PM (eds) Causes of photooxidative stress and amelioration of defense systems in plants. CRC Press, Boca Raton, pp 177–197

Drerup MM, Schlücking K, Hashimoto K, Manishankar P, Steinhorst L, Kuchitsu K (2013) The Calcineurin B-like calcium sensors CBL1 and CBL9 together with their interacting protein kinase CIPK26 regulate the *Arabidopsis* NADPH oxidase RBOHF. Mol Plant 6:559–569

Dutilleul C, Garmier M, Noctor G, Mathieu CD, Chetrit P, Foyer CH, De paepe R (2003) Leaf mitochondria modulate whole cell redox homoeostasis, set antioxidant capacity and determine stress resistance through altered signaling and diurnal regulation. Plant Cell 15:1212–1226

Eltsner EF (1982) Oxygen activation and oxygen toxicity. Annu Rev Plant Physiol 33:73–96

Eltsner EF (1987) Metabolism of activated oxygen species. In: Davis D (ed) The biochemistry of plants San Diego. Academic Press, pp 253–315

Evans NH, McAinsh MR, Hetherington AM, Knight MR (2005) ROS perception in *Arabidopsis thaliana*: the ozone-induced calcium response. Plant J 41:615–626. https://doi.org/10.1111/j.1365-313X.2004.02325.x

Fadzillah NM, Gill V, Flinch RP, Burdon RH (1996) Chilling, oxidative stress and antioxidative response in shoot cultivars of rice. Planta 199:552–556

Forman HJ, Boveris A (1982) Free radicals in biology and medicine, vol 5. Academic Press, New York, pp 65–90

Foyer CH (1997) Oxygen metabolism and electron transport in in photosynthesis. In: Scandalios J (ed) Molcular biology of free radical scavenging enzymes. Cold Spring Harbor Laboratory, New York

Foyer CH, Harbinson J (1994) Oxygen metabolism and regulation of photoelectron transport. In: Foyer CH, Mullineauex PM (eds) Causes of photooxidative stress and amelioration of defense systems in plants. CRC Press, Boca Raton, pp 01–13

Foyer CH, Noctor G (2003) Redox sensing and signaling associated with reactive oxygen in chloroplast, peroxysome and mitochondria. Physiol Plant 119:355–364

Foyer CH, Shigeoka S (2011) Understanding oxidative stress and antioxidant functions to enhance photosynthesis. Plant Physiol 155:93–100

Fridovich I (1986) Superoxide dismutases. Adv Enzymol Relat Areas Mol Biol 58:61–97

Fridovich I (1995) Superoxide and superoxide dismutase. Ann Rev Biochem 64:97–112

Gabig TG (1983) NADPH-dependent superoxide generating oxidase from human nutrophils. J Biol Chem 258:6352–6356

Gill SS, Khan NA, Anjum NA, Tuteja N (2011) Amelioration of cadmium stress in crop plants by nutrients management: morphological, physiological and biochemical aspects. In: Anjum NA, Lopez-Lauri F (eds) Plant nutrition and abiotic stress tolerance III, plant stress 5 (Special Issue 1). Global Science Books Ltd.), Ikenobe, pp 1–23

Gilroy S, Suzuki N, Miller G, Choi WG, Toyota M, Devireddy AR, Mittler R (2014) A tidal wave of signals: calcium and ROS at the forefront of rapid systemic signaling. Trends Plant Sci 19:623–630

Grossman S, Leshem YY (1978) Lowering endogenous LOX activities in *Pisum sativum* by cytokinin as related to senescence. Physiol Plant 43:359–362

Halliwell B (2006) Reactive species and antioxidants: Redox biology is a fundamental theme of aerobic life. Plant Physiol 141:312–322

Halliwell B, Gutteridge JMC (1984) Oxygen toxicity and oxyradicals, transition metals and disease. Biochem J 219:01–19

Halliwell B, Gutteridge JMC (1999) Free radicals in biology and medicine, 3rd edn. Oxford University Press, New York

Hameed A, Goher M, Iqbal N (2013) Drought induced programmed cell death and associated changes in antioxidants, proteases, and lipid peroxidation in wheat leaves. Biol Plant 57:370–374

Hammond-Kosack KE, Jones JDC (2000) Responses to plant pathogen. In: Buchanon BB, Grussem W, Jones R (eds) Biochemistry and molecular biology of plants. American Society of Plant Physiology, Rockville, pp 102–106

Han D, Ybanez MD, Ahmadi S, Yeh K, Kaplowitz N (2009) Redox regulation of tumor necrosis factor signaling. Antioxid Redox Signal 11:2245–2263

Hendry GAF, Baker AJM, Ewart CF (1992) Cadmium tolerance and toxicity, oxygen radical processes and molecular damages in cadmium tolerant and sensitive clones of *Holcuslanatus*. Acta Bot Mecrl 41:271–281

Hernández JA, Jiménez A, Mullineaux P, Sevilla F (2000) Tolerance of pea (*Pisum sativum* L.) to long-term salt stress is associated with induction of antioxidant defenses. Plant Cell Environ 23:853–862doi. https://doi.org/10.1046/j.1365-3040.2000.00602.x

Hideg E, Kalai T, Hideg K, Vass L (1998) Photoinhibition of photosynthesis in vivo results in singlet oxygen production. Detection via nitroxide induced fluorescence quenching in broad bean leaves. Biochemistry 37:11405–11411

Hideg E, Barta C, Kalai T, Hideg K, Vass L, Asada K (2002) Detection of singlet oxygen and superoxide with fluorescent sensors in leaves under stress by photoinhibition or UV radiation. Plant Cell Physiol 11:54–64

Higuchi T (2006) Look back over the studies of lignin biochemistry. J Wood Sci 52:2. https://doi.org/10.1007/s10086-005-0790-z

Imlay JA (2008) Cellular defenses against superoxide and hydrogen peroxide. Annu Rev Biochem 77:755–776

Jabs T (1999) Reactive oxygen species intermediates as mediators of programmed cell eath in plants and animals. Biochem Pharmacol 57:231–245

Jiang M, Zhang J (2001) Effect of ABA on active oxygen species, antioxidative defense system and oxidative damage in leaves of maize seedlings. Plant Cell Physiol 42(11):1265–1273

Kimura S, Kaya H, Kawarazaki T, Hiraoka G, Senzaki E, Michikawa M (2012) Protein phosphorylation is a prerequisite for the Ca^{2+}-dependent activation of *Arabidopsis* NADPH oxidases and may function as a trigger for the positive feedback regulation of Ca^{2+} and reactive oxygen species. Biochim Biophys Acta 1823:398–405. https://doi.org/10.1016/j.bbamcr.2011.09.011

Krieger-Liszkay A (2005) Singlet oxygen production in photosynthesis. J Exp Bot 56:337–346. https://doi.org/10.1093/jxb/erh237

Kwak JM, Mori IC, Pei ZM, Leonhardt N, Torres MA, Dangl JL (2003) NADPH oxidase Atrboh D and Atrboh F genes function in ROS-dependent ABA signaling in *Arabidopsis*. EMBO J 22:2623–2633. https://doi.org/10.1093/emboj/cdg277

Laloi C, Apel K, Danon A (2004) Reactive oxygen signaling: the latest news. Curr Opin Plant Biol 7:323–328

Lamb C, Dixon RA (1997) The oxidative burst in plant disease resistance. Annu Rev Plant Physiol Plant Mol Biol 48:251–275

Legendre L, Rueter S, Heinstei PF, Low PS (1993) Characterization of the oligogacturroxide-induced oxidative burst in cultured soybean cells. Plant Physiol 102:233–240

Levine A (1999) Oxidative stress as regulator environmental responses in plants. In: Ferver HR (ed) Plant responses to environmental stress. Mervel Decker Inc, Switzerland, pp 146–163

Levine A, Tenhaken R, Dixon R, Lamb CJ (1994) Hydrogen peroxide from oxidative burst orchestrates the plant hypersensitive disease resistance response. Cell 79:583–593

Levine A, Pennell RI, Alvarez MF, Palmer R, Lamb C (1996) Calcium mediated apoptosis in a plant hypersensitive disease response. Curr Biol 04:427–437

Lin F, Ding H, Wang J, Zhang H, Zhang A, Zhang Y (2009) Positive feedback regulation of maize NADPH oxidase by mitogen-activated protein kinase cascade in abscisic acid signalling. J Exp Bot 60:3221–3238. https://doi.org/10.1093/jxb/erp157

Mahalingam R, Fredroff N (2003) Stress response, cell death and signaling: the many faces of ROS. Physiol Plant 119:56–68

Maxwell DP, Wang Y, McIntosh L (1999) Alternative oxidase lowers mitochondrial ROS production in plant cells. Proc Natl Acad Sci 96:8271–8276

Maxwell DP, Nickels R, McIntosch L (2002) Evidence of mitochondrial involvement in transduction of signals required for induction of genes associated with pathogen attack and senescence. Plant J 29:269–279

Mehlar AH (1951) Studies on reactions of Illuminated chloroplasts. II Stimulation and inhibition of reaction b with oxygen. Arch Biochem Biophys 33:65–77

Millar AH, Leaver CJ (2000) The cytotoxic lipid peroxidation product, 4-hydroxyl-2-nonenal specifically inhibits dehydrogenase in matrix of plant mitochondria. FEBS Lett 481:117–121

Miller G, Shulaev V, Mittler R (2008) Reactive oxygen signaling and abiotic stress. Physiol Plant 133(3):481–489

Miller G, Schlauch K, Tam R, Cortes D, Torres MA (2009) The plant NADPH oxidase RbohD mediates rapid, systemic signaling in response to diverse stimuli. Sci Signal 2(84):45–49

Miller G, Suzuki N, Ciftci-Yilmaz S, Mittler R (2010) Reactive oxygen species homeostasis and signaling during drought and salinity stresses. Plant Cell Environ 33:453–467

Mittler R (2017) ROS are good. Trends Plant Sci 22(1):11–19

Mittler R, Blumwald E (2015) The roles of ROS and ABA in systemic acquired acclimation. Plant Cell 27:64–70. https://doi.org/10.1105/tpc.114.133090

Moller IM, Jensen PE, Hansson A (2007) Oxidative modifications of cellular proteins. Annu Rev Plant Biol 58:459–481

Murphy MP (2009) How mitochondria produce reactive oxygen species. Biochem J 417(1):1–13. https://doi.org/10.1042/BJ20081386

Neill S, Desikan R, Hancock J (2002) Hydrogen peroxide signalling. Curr Opin Plant Biol:5388–5395. https://doi.org/10.1016/S1369-5266(02)00282-0

Nixon PJ (2000) Chlororespiration. Philos Trans R Soc London 355:1541–1547

Noctor G, Veljovic-Jovanovic SD, Driscoll S, Novitskaya L, Foyer CH (2002) Drought and oxidative load in wheat leaves. A predominant role for photorespiration? Ann Bot 89:841–850

Ogawa K, Kanematsu S, Asada K (1997) Generation of superoxide anion and location of Cu-Zn SOD in the vascular tissues of spinach hypodermis and their association with lignification. Plant Cell Physiol 38:1118–1123

Overmyer K, Brosché M, Kangaskärvi J (2003) Reactive oxygen species and hormonal control of cell death. Trends Plant Sci 8:335–342. https://doi.org/10.1016/S1360-1385(03)00135-3

Pei ZM, Murata Y, Benning G, Thomine S, Klüsener B, Allen GJ, Grill E, Schroeder JI (2000) Calcium channels activated by hydrogen peroxide mediate abscisic acid signalling in guard cells. Nature 406:731–734

Pinto E, Sigaud-kutner T, Leitao MA, Okamoto OK, Morse D, Colepicolo P (2003) Heavy metal-induced oxidative stress in algae. J Phycol 39:1008–1018. https://doi.org/10.1111/j.0022-3646.2003.02-193.x
Porter NA, Caldwell SE, Mills KA (1995) Mechanisms of free radical oxidation of unsaturated lipids. Lipids 30:277–290. https://doi.org/10.1007/BF02536034
Prasad TK, Anderson MD, Martin B, Stewart CR (1994) Evidence of chilling induced oxidative stress and a regulatory role of hydrogen peroxide. Plant Cell 06:65–74
Rao MV, Paliyath G, Ormrod DP (1996) Ultraviolet-B- and ozone-induced biochemical changes in antioxidant enzymes of Arabidopsis thaliana. Plant Physiol 110:125–136
Rhoads DM, Umbach AL, Subbaiah CC, Siedow JN (2006) Mitochondrial reactive oxygen species – contribution to oxidative stress and interorganellar signaling. Plant Physiol 141:357–366
Robson CA, Vanlerberghe GC (2002) Transgenic plant cells lacking mitochondrial alternative oxidase have increased susceptibility to mitochondria –dependent and independent pathways of cell death. Plant Physiol 129:1908–1920
Rouhier N, Stephane D, Lenaire JPJ (2008) The role of glutathione in photosynthetic organism: emerging function of glutaredoxin and glutathionylation. Annu Rev Plant Biol 59:143–166
Sagi M, Fluhr R (2001) Superoxide production by plant homologues of the gp91(phox) NADPH oxidase: modulation of activity by calcium and by tobacco mosaic virus infection. Plant Physiol 126:1281–1290
Scandalios JG (2005) Oxidative stress: molecular perception and transduction of signals triggering antioxidant gene defenses. Braz J Med Biol Res 38:995–1014
Segal AW, Abo A (1993) The biochemical basis of NADPH oxidase of phagocytes. Trends Biochem Sci 18:43–47
Sharma P, Jha AB, Dubey RS, Pessarakli M (2012) Reactive oxygen species, oxidative damage, and antioxidative defense mechanism in plants under stressful conditions. J Bot:1–26. https://doi.org/10.1155/2012/217037
Sharma R, Priya P, Jain M (2013) Modified expression of an auxin-responsive rice CC-type glutaredoxin gene affects multiple abiotic stress responses. Planta 238:871–884
Sochor J, Babula P, Adam V, Krska B, Kizek R (2012) Sharka: the past, the present and the future. Viruses 4:2853–2901
Somashekaraiah BV, Padmaja K, Prasad ARK (1992) Phytotoxicity of cadmium ions on germinating seedlings of mungbean (*Phaseolus vulgaris*): involvement of lipid peroxides in chlorophyll degradation. Physiol Plant 85:85–89
Spiteller G (2003) The relationship between cell wall, lipid peroxidation, proliferation, senescence and cell death. Physiol Plant 119:05–18
Spreitzer H, Schmidt J, Spiteller G (1989) Comparative analyses of the fatty acid fraction of vegetables in dependence on preliminary treatment. Fett Wiss Technol 91:108–113
Strid A, Chow WS, Anderson JM (1994) UV-B damage and protection at the molecular level in plants. Photosynth Res 39:475–489
Suzuki N, Koussevitzky S, Mittler R, Miller G (2011) ROS and redox signaling in response to abiotic stress. Plant Cell Environ. https://doi.org/10.1111/j.1365-3040.2011.02336.x
Suzuki N, Miller G, Salazar C, Mondal HA, Shulaev E, Cortes DF (2013) Temporal-spatial interaction between reactive oxygen species and abscisic acid regulates rapid systemic acclimation in plants. Plant Cell 25:3553–3569. https://doi.org/10.1105/tpc.113.114595
Thompson JE, Legge RE, Barber RF (1987) Role of free radicals in senescence and wounding. New Phytol 105:313–344
Torres MA, Dangl JL, Jones JDG (2002) *Arabidopsis* $gp^{91\ phox}$ homologues atrboh D atrboh F are required for acclimation reactive oxygen intermediates in plant dependent response. Proc Natl Acad Sci U S A 99:517–522
Van Assche F, Clijsters H (1990) Effect of metals on enzymic activity in plants. Plant Cell Environ 13:195–206
Vanova E, VanBreusegerm F, Dat J, Belles-Bolx E, Inze D (2002) Role of reactive oxygen species in signal transduction. In: Scheel D, Wasternack C (eds) Plant signal transduction. Oxford University Press, Oxford, pp 41–73

Winston GW (1990) Physicochemical basis of free radical formation in cells: production and defenses. In: Smallwood W (ed) Stress responses in plants: adaptation and acclimation mechanisms. Willy-Liss Inc, New York, pp 57–86

Wiseman H, Halliwell B (1996) Damage to DNA by reactive oxygen species and nitrogen species. Cancer Biochem J 387:856–870

Zhou YH, Yu JQ, Mao WH, Huang LF, Song XS, Nogués S (2006) Genotypic variation on Rubisco expression, photosynthetic electron flow and antioxidant metabolism in the chloroplasts of chill-exposed cucumber plants. Plant Cell Physiol 47:192–199

ROS and Antioxidants: Relationship in Green Cells

2

Abstract

The redox homeostasis of plant cell, which largely depends on prooxidant and antioxidant status, is perturbed under environmental assault. In fact, the imposition of abiotic and biotic stresses changes redox status or homeostasis of the plant cell toward prooxidants and leads to a condition called oxidative stress. Orchestrated antioxidative defense that largely comprises of information-rich redox buffers and enzymes ensues to combat the situation, specifically at the site of action of the stress. Thus, the functional roles of these antioxidative defense responses include the restoration of metabolic redox homeostasis, the protection of the photosynthetic machinery, the preservation of membrane integrity, the protection of nucleic acids and proteins, etc. Current progress of work suggests that complex regulatory mechanisms function at both the gene expression and protein level to coordinate antioxidant responses in plants. To ensure survival, particularly under stress, plants have developed both enzymatic and nonenzymatic defense systems which work hand in hand not only to scavenge ROS but also to tightly regulate the endogenous titer of ROS to induce the signaling role of ROS, required for stress acclimation. Growing body of evidence suggests the role of redox homeostasis under environmental stress in which ROS-antioxidant interaction acts at metabolic interface for signal derived from unfavorable environmental cues. The current research also suggest the function of antioxidants as key arbitrator of intracellular redox status and exhibit high degree of potential with their differential cellular status permitting antioxidant-driven vectorial signaling. In this chapter, an effort has been made to detail the different antioxidative defense mechanisms operating in the cellular and subcellular level for scavenging ROS and maintaining redox homeostasis under stressful conditions. Further, a descriptive account regarding antioxidant-based redox information which influences gene expression associated with unfavorable environmental conditions to maximize defense is also discussed.

S. Bhattacharjee, *Reactive Oxygen Species in Plant Biology*,
https://doi.org/10.1007/978-81-322-3941-3_2

Keywords
Antioxidants · Redox buffer · Antioxidative enzymes · ROS scavengers · Antioxidant signaling

2.1 Introduction

Among several factors, the frail but significant equilibrium between generation of ROS and their scavenging determines largely the redox fate of the cell, i.e., whether ROS will impose oxidative damage or induce redox signaling. Therefore, it is necessary for the cells to control the ROS turnover or tightly regulate its endogenous titer so that it can avoid oxidative injury and exploit its signaling role efficiently. Scavenging of excess ROS for the maintenance of functional oxidant level is achieved by a well-organized antioxidative defense system comprising of both nonenzymatic and enzymatic antioxidants (Mittler 2017; Das and Roychoudhuri 2014; Foyer and Noctor 2005a; Foyer and Shigeoka 2011). Superoxide dismutase (SOD), catalase (CAT), guaiacol peroxidase (GPX), and enzymes of ascorbate-glutathione (AsA-GSH) cycle such as ascorbate peroxidase (APX), monodehydroascorbate reductase (MDHAR), dehydroascorbate reductase (DHAR), and glutathione reductase (GR) are the basic enzymatic components for redox-detoxification (Das and Roychoudhuri 2014; Foyer and Noctor 2005a; Foyer and Shigeoka 2011). Low molecular weight compounds like ascorbate (AsA), glutathione (GSH), carotenoids, and tocopherols and secondary metabolite phenolic acids and flavonoids serve as potent nonenzymatic antioxidants. The process associated with ROS production, coupled with its removal by an efficient antioxidative defense system, occurs constantly in cells to prevent the potential toxic effect of ROS that could damage all the major macromolecules of the cell, collectively referred to as oxidative stress. Several workers have reported a significant upregulation in activities of antioxidant defense system in plants to combat oxidative stress. Maintenance of upregulated antioxidant defense capacity to quench and scavenge the toxic level of ROS has been linked to increased tolerance of plants to environmental stresses (Foyer and Noctor 2005a, b; Asada 2006; Mittler 2017). Recent time witnessed significant progress of work in improving stress-induced (both abiotic and biotic) oxidative stress tolerance in crop plants by developing transgenic lines with enhanced levels of antioxidants and quenchers (Mittler 2017; Foyer and Shigeoka 2011). In this context, expressing multiple antioxidant enzymes or combined defense system proved to be more efficient than individual expression for antioxidants in transgenic plants with enhanced tolerance to multiple environmental stresses (Foyer and Shigeoka2011). The present chapter focuses on types and roles of antioxidative defense system and their regulation in combating danger posed by excessive generation of ROS. Moreover, the role of antioxidants in redox sensing is also a subject of discussion.

2.2 Antioxidant Defense Machinery of Plant Cell

Cellular maintenance of redox homeostasis largely depends on antioxidant defense mechanism which helps not only to alleviate the oxidative damages but also act as redox buffer necessary for the maintenance of physiological level of ROS for signaling. In general, the cellular antioxidant mechanism has two defense weapons with the enzymatic components and nonenzymatic antioxidants, which work alone or in concerted manner for detox-scavenging and/or mitigation of genesis of ROS (described in Table 1.1).

2.2.1 Enzymatic Antioxidants

The antioxidant defense enzymes confined to the different subcellular locales of plant cell that work in isolated way or in tandem include superoxide dismutase (SOD, E.C.1.15.1.1), catalase (CAT, EC1.11.1.6), ascorbate peroxidase (APX, EC1.11.1.11), monodehydroascorbate reductase (MDHAR, EC1.6.5.4), dehydroascorbate reductase (DHAR, EC1.8.5.1), glutathione reductase (GR, EC1.8.1.7), and guaiacol peroxidase (GPX, EC1.11.1.7). Apart from this plant also exploit typical antioxidant system, where several low molecular antioxidants along with antioxidative defense enzymes work in tandem to detoxify ROS needed for restoring redox homeostasis (ascorbate-glutathione pathway, ascorbate-tocopherol defense system, etc.).

2.2.1.1 Superoxide Dismutase (SOD)

The enzyme superoxide dismutase (SOD,E.C.1.15.1.1) is an omnipresent metalloenzyme and the first line of defense against ROS-induced oxidative damages. The SOD catalyzes dismutation of superoxide to O_2 and H_2O_2. This reaction, in fact, removes the prospect of $OH^{\bullet}$ generation by the Haber-Weiss reaction. In plants, three isoforms of SODs exist, based on the metallocofactors it binds. These are mitochondrial Mn-SOD, chloroplastic Fe-SOD, and cytosolic, peroxisomal, or chloroplastic Cu/ZnSOD (Mittler 2002). The activity of SOD has been found to be upregulated by unfavorable environmental conditions (Boguszewska et al. 2010).

$$O_2^{\cdot-} + O_2^{\cdot-} + 2H^+ \rightarrow SOD \rightarrow 2H_2O + O_2$$

Transgenic expressing higher level of SOD under higher salt or drought stress plays critical role in the survival of plants under such condition (Badawi et al. 2004). SOD made the first line of defense against ROS. Three different isoenzymes of SOD, namely, Cu/ZnSOD (dimmers, found in cytosolic fraction and also in chloroplast in higher plant), Mn-SOD enzymes (tetramers, found in the mitochondria and peroxisome), and the Fe-SOD (present in chloroplasts), are found in plant cell. The upregulation in the activity of SOD occurs to mitigate oxidative stress under unfavorable environmental conditions plays a critical role in survival of plants under such situation (Gambarova and Gins 2008, Kukreja et al. 2005, Gapinska et al.

2008). The subcellular distribution of SOD isoenzymes is noticed in plants (Torres 2010). Environmental stress such as salinity exhibited three SOD activity bands (Mn-SOD, Fe-SOD, and Cu/ZnSOD) in *C. arietinum* (Eyidogan and Oz 2005). Pan et al. (2006) have noticed significant increase in the activities of Cu/*ZnSOD* and Mn-SOD isozymes under salt stress. An upregulation in SOD activity following drought stress was noted in *O. sativa* (Sharma and Dubey 2005a), *P. vulgaris* (Zlatev et al. 2006), and *Alternanthera philoxeroides* (Wang et al. 2008). Co-occurrence of drought and excess photochemical energy significantly augmented the SOD activity in comparison to corresponding individual stresses. Transgenic rice plants overexpressing OsMT1 exhibited better drought tolerance (Yang et al. 2009). In *Arabidopsis* and *L. esculentum,* the overexpression of a Mn-SOD also exhibited salt tolerance (Wang et al. 2004, 2007).

2.2.1.2 Catalase (CAT)

Another antioxidative enzyme, CAT (E.C.1.11.1.6), responsible for catalyzing the dismutation of H_2O_2 into H_2O and O_2, is a heme-containing tetrameric enzyme. It exhibited significantly high affinity for hydrogen peroxide (H_2O_2), but lesser kinetic specificity for other organic hydroperoxides (R-O-O-R). It shows a significantly high turnover rate (6×10^6 molecules of H_2O_2 to H_2O and O_2 min^{-1}). It is unique among H_2O_2-degrading enzymes as it doesn't need any reducing equivalent for the reaction. Peroxisomes, which are the hotspots of H_2O_2 production, often suffer oxidative stress, as the organelle sponsors photorespiration, β-oxidation of fatty acids, and purine metabolism (Mittler 2002). Peroxisomes for this reason exhibit greater kinetics of CAT activity. The recent reports also suggest that CAT is also associated with other subcellular parts such as the cytosol, chloroplast, and the mitochondria, though significant CAT activity, as that of peroxisome, is yet to be seen (Mhamdi 2010). Higher plants in general exhibited the presence of three *CAT* genes. *CAT1* of these multigene families is primarily expressed in seeds and pollens and is peroxisomal and cytosolic. On the contrary, *CAT2* is found to be predominantly expressed in chlorophyllous tissues and also in roots and seeds. They are primarily peroxisomal and cytosolic. The third form of catalase, *CAT3*, is found to be expressed in living cells of vascular tissues and leaves and is mitochondrial.

CAT removes and detoxifies the H_2O_2 in an energy-efficient way by sponsoring the following reaction:

$$H_2O_2 \rightarrow H_2O + (1/2)O_2$$

The structural identity of the enzyme CAT is well characterized (Mhamdi 2010). It is a heme-containing peptide that catalyzes the dismutation of H_2O_2 into H_2O and O_2 without using a coreactant. It is ubiquitous, and its function is not only to remove the H_2O_2 generated in peroxisomes by β-oxidation of fatty acids, C2 cycle, and purine metabolism but also to regulate the endogenous titer of the oxidant (Mittler 2002; Vellosillo et al. 2010). The significance of the enzyme can be vouched by the proliferation of peroxisomes during acute stresses, which might help in regulating the endogenous titer of H_2O_2 that affords cell protection by redox signaling

(Lopez-Huertas et al. 2000, Bhattacharjee 2012). Catalase is also a kinetically very efficient enzyme which is capable of sponsoring the highest turnover rates among the antioxidative defense enzymes: one molecule of CAT in 1 min can break down about six million H_2O_2 molecules to H_2O and O_2. Environmental stress that reduces the rate of protein turnover, such as extremes of temperature, salinity, and drought, was also found to reduce CAT activity (Karuppanapandian et al. 2006a, b; Karuppanapandian and Manoharan 2008; Chen et al. 2010; Hojati et al. 2010). Catalase is also a light-sensitive protein that has a high rate of turnover under environmental stresses (Boguszewska et al. 2010, Mhamdi 2010). There exists contradictory report on CAT activity under stress. In *C. arietinum, roots* exposed to salinity stress cause an upregulation in its activity, whereas in the other study, a decrease in CAT activity in rice seedling under drought stress was observed (Kukreja et al. 2005; Sharma and Dubey 2005a, b). *O. sativa* transgenic showed upregulation of CAT activity by the introduction of katE gene under salt stress (Nagamiya et al. 2007). In a similar study, Eyidogan and Oz (2005) and Kukreja et al. (2005) observed a significant upregulation in CAT activity in *C. arietinum* leaves and roots under salinity stress. Srivastava et al. (2005), on the contrary, reported a decrease in CAT activity in *A. doliolum* under NaCl salinity stress, hinting susceptibility of the plant to stress. Simova-Stoilova et al. (2010) reported differential activity of CAT in wheat cultivars showing different extents of drought stress resistance. The excessive stress downregulates the activity of CAT that could be vouched from the study of Pan et al. (2006) who have noticed that the combined effect of high magnitude of salt and drought stress reduces the activity of CAT in *Glycyrrhiza uralensis* seedlings. Transgenesis for catalase gene to the susceptible variety on the other hand enhances tolerance to the crop (Yang et al. 2009). Therefore, the role of the first line of antioxidative defense enzyme, the catalase, particularly under stress, is well established and needs to be understood not merely as an ordinary antioxidative enzyme but as modulator of oxidative stress for initiating redox signaling (Bhattacharjee 2012; Mittler 2017).

2.2.1.3 Ascorbate Peroxidase (APX)

One of the peroxidase categories of enzyme, APX (APX, E.C.1.1.11.1), which works as a vital constituent of the ascorbate-glutathione (ASC-GSH) cycle degrades H_2O_2 to H_2O just like CAT with a great degree of differences when compared. While CAT predominantly evokes its action in the peroxisomes, APX does it in the chloroplast and cytosol. CAT works alone in breaking H_2O_2, whereas APX reduces H_2O_2 to H_2O and DHA (dehydroascorbate) using ascorbic acid (AA) as a reductant.

$$H_2O_2 + \text{Ascorbate} \rightarrow 2H_2O + \text{DHA}$$

Five different isoforms of APX based on different amino acid sequences and sites of their presence, viz., cytosolic, mitochondrial, peroxisomal, stromal, and thylakoidal, have been noticed (Sharma and Dubey 2004). Because of its (APX) wider distribution and significantly better affinity for the substrate H_2O_2 than CAT, it is a more efficient scavenger of H_2O_2, primarily under stress.

The role of APX under stress in the maintenance of redox homeostasis and protection of cells in higher plants is a well-recognized phenomenon. Precisely, in green cell, during photosynthetic operation under stress, APX scavenges excess H_2O_2 in water-water and ascorbate-glutathione cycles (Halliwell-Asada cycle) and utilizes ascorbate as the coreactant electron donor. In Halliwell-Asada cycle, APX capitalizes ascorbate as a hydrogen donor and converts H_2O_2 to monodehydroascorbate (MDHA) and H_2O (Asada 2000). $O_2^{\cdot -}$ generated may be converted to H_2O_2 by SOD, which is subsequently scavenged by membrane-bound APX (Asada 2000, 2006). Like SOD, the APX family consists of at least five different isoforms, namely, cytosolic APX (cAPX), thylakoid APX (tAPX), glyoxisomal APX (gAPX), membrane APX (mAPX), and the chloroplast stromal soluble APX (sAPX) (Noctor and Foyer 1998a, b). Enhanced activity of APX in general is found to be associated with environmental stress (Srivastava et al. 2005). Drought stress in general upregulates chloroplastic-APX activity and subsequently reduces their activity under extreme drought condition (Sharma and Dubey 2005a, b). Koussevitzky et al. (2008) showed that cytosolic APX1 plays an extremely important role in the protection of plants under synergistic drought and heat stress. Badawi et al. (2004) showed that the overexpression of APX gene in *Nicotiana tabacum* chloroplasts enhanced plant tolerance to salt and dehydration stress. Yang et al. (2009) correlated with the enhanced tolerance of OsMT1a overexpressing transgenic rice under drought. In a study related to the expression patterns of different isoforms of APX in roots of etiolated *O. sativa* seedlings under salinity stress, the transcript abundance for two cytosolic APXs (OsAPX1 and OsAPX2), two peroxisomal APXs (OsAPX3 and OsAPX4), and four chloroplastic APXs (OsAPX5, OsAPX6, OsAPX7, and OsAPX8) showed significant variations. It was noted that salinity stress imposed to the seedlings enhances the expression of OsAPX8 and activities of APX, but on the contrary, the effect on the expression of other isoforms of the genes of the same enzyme (OsAPX1, OsAPX2, OsAPX3, OsAPX4, OsAPX5, OsAPX6, and OsAPX7) in rice roots grown under salinity stress seems to be insignificant (Hong et al. 2007). Overexpression of OsAPXa or OsAPXb in *Arabidopsis* exhibited increased salt tolerance. It was also noted that the upregulation of OsAPXb caused enhanced activity and maintenance of APX than OsAPXa in transgenic plants under salinity stress (Lu et al. 2007). The transcript level of cytosolic APX gene also showed significant enhancement in the alfalfa nodule experiencing drought stress (Naya et al. 2007). All the experimental data conclusively support the significant role of APOX in the maintenance of redox homeostasis and stress tolerance in plants.

2.2.1.4 Monodehydroascorbate Reductase (MDHAR)

The regeneration of ascorbic acids (AA) from short-lived MDHA (monodehydroascorbic acid), using NADPH as a reducing agent, requires MDHAR (E.C.1.6.5.4). It causes replenishment of the cellular AA pool. As it regenerates AA, it is localized in the same subcellular locale with the APX in the peroxisomes and mitochondria, where APX reduces H_2O_2 with the help of AA in ascorbate-glutathione pathway (Mittler 2002). As other antioxidative enzymes, MDHAR also possess several isozymes associated with mitochondria, peroxisomes, chloroplast, cytosol, and glyoxysomes.

$$\text{MDHA} + \text{NADPH} \rightarrow \text{AA} + \text{NADP}^+$$

In plant, MDHAR is a major antioxidative enzyme of ascorbate-glutathione pathway which protects cell damage against ROS. This enzyme is chloroplast-, cytosol-, mitochondria-, glyoxysome-, and leaf peroxisome-localized isoforms (del Río et al. 2002a, b; Mittler 2002). It is basically a flavin adenine dinucleotide (FAD) enzyme that is mainly present as chloroplast and cytosol. Two enzymes, MDHAR and DHAR, are involved in the regeneration of reduced ascorbate. MDHAR uses NAD(P)H directly to recycle ascorbate. In fact, MDHA is itself an efficient electron acceptor and hence an active component of ascorbate-glutathione pathway (Noctor and Foyer 1998a, b; Asada 2000). That these enzymes, i.e., MDHAR, DHAR, and GR, are involved in the regeneration of ascorbate and work in tandem is evident from their higher activities in drought-stressed rice seedlings (Sharma and Dubey 2005a, b). Transgenic tobacco with overexpression of MDHAR caused increased tolerance against salt and osmotic stresses (Eltayeb et al. 2007). The enzyme also exhibits a high specificity for the substrate MDHA as the electron acceptor, ultimately preferring NADH rather than NADPH as the electron donor. Asada (1999) unfolded the mechanism of the multi-step reduction of FAD, where the first stem commences with the reduction of the enzyme-FAD component to form a charge-transfer complex. The reduced FAD-enzyme then donates electrons to two MDHAs, producing two molecules of ascorbate. The deprotonation by photoreduced ferredoxin (red Fd) in the thylakoids is of great importance. As the reduced Fd can donate electron to MDHA more effectively than NADPH, MDHAR failed to sponsor reduction of MDHA in the thylakoidal ROS scavenging system (Asada 1999). The activities of all the enzymes involved in the regeneration of ASH, i.e., MDHAR, DHAR, and GR, were higher in drought-stressed rice seedlings (Sharma and Dubey 2005a, b, Bhattacharjee and Dey 2017).

2.2.1.5 Dehydroascorbate Reductase (DHAR)

Another AA-regenerating enzyme, DHAR (E.C.1.8.5.1), reduces dehydroascorbate (DHA) to AA using reduced glutathione (GSH) as an electron donor (Eltayeb et al. 2007). As the regulation of the AA pool size in both symplast and apoplast is extremely important from the point of view of redox regulation of cell, the activity of this enzyme is extremely important (Chen and Gallie 2006). DHAR is found to be abundantly localized in seeds, roots, and green and etiolated shoots.

$$\text{DHA} + 2\text{GSH} \rightarrow \text{AA} + \text{GSSG}$$

DHAR, being an integral component of ascorbate-glutathione cycle, helps in the regeneration of AA from its oxidized state. Formation of MDHA is followed by the univalent oxidation of ascorbic acid which then got converted to a divalent oxidation product DHA (dehydroascorbate). DHA is subsequently reduced to AA by DHAR in a reaction with participation of GSH as an electron donor (Eltayeb et al. 2007). The activity of DHAR is found to be linked with stomatal opening and closing and hence seems to play a vital role in water-use efficiency of plants under drought and salinity stress (Chen and Gallie 2005; Ushimaru et al. 2006; Eltayeb et al. 2006)

2.2.1.6 Glutathione Reductase (GR)

The Halliwell-Asada pathway enzyme GR (E.C.1.6.4.2) is basically a flavoprotein oxidoreductase which uses NADPH as the donor of electron to reduce GSSG to GSH. As the reduced form of glutathione (GSH) is used up to regenerate AA either from MDHA or DHA, it is getting converted to its oxidized form (GSSG). It catalyzes the formation of a disulfide bond in glutathione disulfide to maintain redox homeostasis in the form of a high cellular GSH/GSSG ratio. It is predominantly associated with chloroplasts though also noticed in mitochondria and cytosol. GSH can react with detrimental ROS members like $^1O_2$ and $OH^{\bullet}$ and detoxify them.

$$GSSG + NADPH \rightarrow 2GSH + NADP+$$

The enzyme associated with both prokaryotes and eukaryotes is an important component of the Halliwell-Asada pathway and plays an essential role in the defense system against oxidative damages triggered by ROS with the help of tripeptide reductant GSH (Edwards et al. 1990; Creissen et al. 1994). GR catalyzes the regeneration of GSH, an antioxidant molecule involved in many regulatory and antioxidative defense processes in plants where it catalyzes the NADPH-dependent reduction of disulfide bond of GSSG and thus replenishing GSH pool (ChalapathiRao and Reddy 2008). An enhancement of GR activity in the leaf tissue of *C. arietinum* L. cv. Gokce and rice under salt and drought stress was also noticed, corroborating the significance of the enzyme in redox regulation under abiotic stress (Eyidogan and Oz 2005; Kukreja et al. 2005; Bhattacharjee and Dey 2017).

2.2.1.7 Guaiacol Peroxidase (GPX)

Another antioxidative defense enzyme, GPX (E.C.1.11.1.7), is a heme-containing enzyme with 40–50 kDa monomers, which take part in the detoxification of excess H_2O_2 both during normal metabolism and stress. Apart from that, it plays a vital role in the biosynthesis of lignin and defends against biotic stress by degrading indole acetic acid (IAA) with the help of H_2O_2 in the said process. GPX uses the aromatic compounds like guaiacol and pyrogallol as the electron donors in the process. Since GPX is active in both extracellular and intracellular environments, it is considered as the key ROS (**H_2O_2**) removal enzyme (Asada 1999).

Antioxidative enzymes	Reactions
Superoxide dismutase (SOD, EC 1.15.1.1):	**$O^{2\cdot -} + O^{2\cdot -} + 2H+ \rightarrow 2H_2O_2 + O_2$**
Catalase (CAT, EC 1.11.1.6):	**$H_2O_2 \rightarrow H_2O + ½ O_2$**
Ascorbate peroxidase (APX, EC 1.11.1.11):	**H_2O_2 + AA → 2 H_2O + DHA**
Guaiacol peroxidase (GPX, EC 1.11.1.7):	**H_2O_2 + GSH → H_2O + GSSG**
Monodehydroascorbate reductase (MDHAR, EC 1.6.5.4):	**MDHA + NAD(P)H → AA + NAD(P)+**
Dehydroascorbate reductase (DHAR, EC 1.8.5.1):	**DHA + 2GSH → AA + GSSG**

2.2.2 Nonenzymatic Antioxidants and Free Radical Quenchers

2.2.2.1 Ascorbic Acid (AA)

Ascorbic acid, the most abundant and extensively studied antioxidant, is capable of donating electrons to a wide range of enzymatic and nonenzymatic reactions for antioxidant function. AA in plant cells is basically produced by Smirnoff-Wheeler pathway in the plant mitochondria, under the catalytic influence of L-galactono-γ-lactone dehydrogenase. The other source of AA is D-galacturonic acid. Both cytosol and apoplast are considered as the storehouse of AA, thus making it the initial line of defense against oxidative stress (Barnes et al. 2002). AA metabolism primarily involves its oxidation through two successive steps, starting with oxidation into MDHA, which if not reduced straight away to ascorbate, might disproportionate to DHA and AA. It got the capacity to react with ROS like H_2O_2, $OH^{\bullet}$, $O^{\bullet -}{}_2$, and tocopheroxyl radical, thereby defending the membranes from ROS-induced oxidative damage (Shao et al. 2005). It also protects the activities of metal-binding enzymes, which are susceptible to oxidative stress. AA plays an important role in xanthophyll cycle by being the cofactor of violaxanthin de-epoxidase and protects chloroplast by dissipation of the excess photochemical energy (Smirnoff 2000). The photo-protective role of AA can be vouched by its involvement in the prevention of photooxidation caused by pH-mediated change of PSII activity and its downregulation, associated with zeaxanthin formation.

Ascorbic acid is one of the most powerful and abundant antioxidants in most plant cell types, organelles, and apoplast (Horemans et al. 2000; Sminoff 2000). Its concentration varies with the developmental stages and found to be maximum in mature leaves having fully developed chloroplast and maximum chlorophyll content. Among the subcellular localizations, maximum accumulation of the ascorbate (30–40% of the total ascorbate) is being noticed in the chloroplast. The stomatal concentrations of ascorbate are found to be as high as 50 mM. In cellular environment, AA got the capacity to reduce a wide range of enzymatic and nonenzymatic molecules, including radical species by donating electrons. AA can reduce and detoxify ROS like $O_2^{\cdot -}$, $OH\cdot$, $^1O_2$, and H_2O_2 to H_2O mainly via APX reaction (Noctor and Foyer 1998a, b). Regeneration of α-tocopherol from its oxidized state of tocopheroxyl radical depends on AA and hence provides membrane stability (Horemans et al. 2000; Smirnoff 2000). Oxidation of AA produces MDHA, which if not reduced to form ascorbates gets subsequently disproportionated into AA and DHA.

2.2.2.2 Reduced Glutathione (GSH)

The water-soluble tripeptide glutathione (γ-glutamyl cysteine glycine) is a low molecular weight thiol-based antioxidant compound, abundantly found in chloroplasts, cytosol, ER, peroxisomes, mitochondria, vacuoles, and apoplast. Its function ranges from cell differentiation, cell growth/division, cell death, and senescence to regulation of metabolism and nutrient transport, detoxification of xenobiotics, synthesis of proteins and nucleotides, synthesis of heavy metal-binding phytochelatins, and finally expression of stress-responsive genes (Mullineaux and Rausch 2005).

This efficiency and versatility of GSH as antioxidant is primarily due to its elevated reductive potential. In fact, a central cysteine residue having nucleophilic character is the cause of its reducing power. GSH efficiently scavenges ROS like $O^{\bullet-}_2$, H_2O_2, $OH^{\bullet}$, and $^1O_2$ and protects almost all important biomolecules by reducing them in the presence of ROS or by forming organic free radical adducts (glutathiolated) thereby generating GSSG as a by-product. GSH also plays a very important role in Halliwell-Asada pathway by regenerating AA. The GSSG thus generated in the said process is converted back to GSH by GR or by its de novo synthesis. This ultimately replenishes the total pool of cellular GSH, which is absolutely fundamental in redox regulation of cell. GSH is also a precursor of phytochelatins, the heavy metal-binding polypeptide, necessary for heavy metal sequestration (RoyChoudhury et al. 2012a). Its ability to chelate heavy metal ions reduces the potential source of ROS formation in plants (RoyChoudhury et al. 2012b). So the maintenance of the delicate balance between GSH and GSSG is a prerequisite for regulating the redox homeostasis of the cell.

2.2.2.3 α-Tocopherol

The lipophilic antioxidant, α-tocopherol, is an efficient scavenger of ROS and secondary oxidative stress products like alkoxy, peroxy, and lipid radicals, making them not only important protectors of biological membranes but also its function (Holländer-Czytko et al. 2005; Kiffin et al. 2006). Among the four isomers (α-, β-, γ-, δ-), the α-tocopherol has the highest antioxidant capability. This class of antioxidant can be synthesized only by photosynthetic organisms, from γ-tocopherol by γ-tocopherol-methyl-transferase (γ-TMT encoded by *VTE4*). Tocopherols are known for their ability to protect galactolipids and other membrane constituents of the chloroplasts by reacting with ROS like $^1O_2$ and $O_2^{\bullet-}$ and quenching their surplus energy, thus conferring the photo-protective role through prevention of photodynamic damages of PSII. Tocopherol has significant contribution in downregulating the membrane lipid peroxidation (MLPO) by serving as an effective free radical sink or trap by halting the chain propagation event of the MLPO cycle. It is capable of quenching excess energy of lipid radicals ($RO^{\bullet}$, $ROO^{\bullet}$, and $RO^{\bullet}$) at the membrane-water interface itself being converted into tocopherol radical ($TOH^{\bullet}$). Subsequently, the $TOH^{\bullet}$ radical gets back to its reduced form by interacting with other antioxidants of the system like AA or GSH (Igamberdiev et al. 2004).

2.2.2.4 Carotenoids

Another lipophilic antioxidants, localized in the plastids of both photosynthetic and non-photosynthetic plant tissues, the carotenoids, exhibit their antioxidative defense property in protecting chloroplasts against oxidative stress by sponsoring the following mechanisms:

(a) Trapping the LPO products to end the chain reactions of MLPO.
(b) Scavenging $^1O_2$ generated in PSII under excess photochemical energy, generating heat as a by-product.

(c) Preventing the generation of highly toxic $^1O_2$ by reacting with $^3Chl^*$ and excited chlorophyll (Chl*).
(d) Dissipating the excess excitation energy of antenna systems of PSII and I, via the xanthophyll cycle (Mortensen et al. 2001).

As antenna molecules, carotenoids can absorb 450–570 nm of the visible spectrum and pass the captured energy to the reaction center during operational Z-scheme.

2.2.2.5 Phenolics

Out of several biochemical properties that plant polyphenols exhibit, the most notable and significant one is their capacity to work as antioxidant. The antioxidant properties of plant polyphenols particularly phenolic acids and flavonoids depend upon the presence of functional groups and their arrangements about the nuclear structure. The configuration and the availability of -OH group significantly influence the mechanism of antioxidant property, radical scavenging ability, and metal chelating ability (Pandey et al. 2012). The configuration of OH groups in the ring "B" of the flavonoids is the most significant determinant of antioxidant property (Pandey et al. 2012). The reason for the significance of "B" ring-associated -OH group configuration in governing the antioxidant property is due to their ability to donate electrons or hydrogen to reactive oxygen species like OH radicals, peroxyl radical, alkoxy radical, etc., thereby stabilizing them and simultaneously giving rise to stable flavonoid radicals (Cao et al. 1997). Flavonoid action in most of the cases involves either suppression of ROS formation by chelating transition metal ions or inhibition of enzymes associated with the formation of ROS. Flavonoids also got the capacity to scavenge ROS or to upregulate antioxidative defense system (Halliwell and Gutteridge 1998; Mishra et al. 2013; Fini et al. 2011). The enzyme microsomal monoxidases, glutathione reductases, glutathione S-transferase, NADH oxidase, and succinate dehydrogenase are in general upregulated by flavonoids (Brown et al. 1998). Peroxidation of membrane lipid is a serious event associated with oxidative damage under severe oxidative stress. Flavonoids, in general, exhibit significant impact in downregulating lipid peroxidation either by chelating transition metal ions (Fe^{2+}/Fe^{3+}) necessary for chain propagation reaction or quenching different radical species (alkoxy, peroxy, lipid radicals) formed during the process. In fact, the radical species generated during lipid peroxidation got the power to induce fresh round of lipid peroxidation. So reducing the free radicals formation by MLPO inhibits nonenzymatic lipid peroxidation (Mishra et al. 2013; Agati et al. 2012). Due to the lower redox potential of flavonoids, these classes of polyphenols are thermodynamically able to reduce the potent ROS formed in the system. The flavonol quercetin in particular got the capacity to chelate iron necessary for propagation of lipid peroxidation. Epicatechin and rutin are strong ROS scavengers and inhibitor of lipid peroxidation (Mishra et al. 2013). The heterocyclic "C" ring of flavonoids also contributes to its antioxidant activity. The free hydroxyl group of third carbon (C3) of the heterocyclic ring permits conjugation with metal ions. The removal of 3-OH exhibits coplanarity and conjugation which compromises the scavenging ability of flavonoids. It often happens that "B" ring OH groups form H-bonds with 3-OH, thereby aligning the "B" ring with the heterocycle and "A" ring. Due to this intramolecular H-bonding, the influence of 3-OH is enhanced by the presence of a 3′, 4′- catechol, thereby enhancing the potent antioxidant activity of flavan-3-ol and flavon-3-ol (Rice-Evans and Burdon 1993).

A significant alteration of membrane fluidity of hydrophilic and hydrophobic moieties may be attributed to this effect suggesting the role of flavonoids in reducing membrane fluidity (Tsuchiya and Linuma 2000).

2.2.2.6 Proline

The amino acid proline, apart from its role as an osmolyte, is also a dominant antioxidant in plant cell. It is widely used to counteract the damaging effects of different ROS members, including the secondary products of oxidative stress. Synthesized from glutamic acid via pyrroline-5-carboxylate (P5C) intermediate, the production of proline seems to be significantly upregulated under stress. Both enzymes, $ð^1$-pyrroline-5-carboxylate synthetase (P5CS) and pyrroline-5-carboxylate reductase (P5CR) necessary for this pathway, exhibit their upregulation under stress. Proline is in fact an efficient scavenger of ROS like $OH^{\bullet}$ and $^1O_2$ and can slow down the damages caused by MLPO. Both the enhanced synthesis and reduced degradation might be responsible for its elevated level under stress (Verbruggen and Hermans 2008). The plant trying to restore the redox homeostasis through synthesis of proline under stress is a well-accepted phenomenon.

2.2.2.7 Mannitol

In response to abiotic stress like dehydration and salinity, mannitol is found to be accumulated in many plant species (Stoop et al. 1996). Though it did not affect photosynthesis, its abundance seems to protect plant cells experiencing dehydration stress. Transgenics of tobacco capable of synthesizing mannitol, by introducing *E.coli* mt1D gene encoding mannitol dehydrogenase, confer drought tolerance. This transgenic tobacco accumulates significantly more amount of mannitol than the salt-tolerant one (Tarczynski et al. 1993). Mannitol is capable of scavenging ROS like hydroxyl radicals, thereby protecting the plants from oxidative damages (Smirnoff 2000). Plants facing oxidative stress can combat the situation by upregulating the formation of mannitol (Biehler and Fock 1996).

2.3 Regulation of Antioxidative Defense System in Plants as a Mechanism to Combat Environmental Stress

Antioxidative defense systems, enzymatic and nonenzymatic, work in tandem with ROS pathway to maintain redox homeostasis in plant cell. Previous works emphasized the importance of antioxidative defense components in regulating redox homeostasis in crop plants (Gaber 2010; Foyer and Noctor 2005a, b). The japonica rice genome possesses several genes necessary for encoding the isoforms of SODs, which includes plastidic SOD (*pCuZnSOD*), cytosolic copper-zinc SODs (*cCuZnSOD1* and *cCuZnSOD2*), one manganese SOD (*Mn-SOD1*), iron SODs (*Fe-SOD2* and *Fe-SOD3*) and putative CuZnSOD-like (*CuZnSOD-L*) (Nath et al. 2014). Overexpressing *Mn-SOD1* in rice exhibited significant reduction in mitochondrial $O_2^{\bullet-}$ formation under stress and subsequently caused reduced stress induction of *OsAOX1a/b* (Li et al. 2013).

Like SODs, in rice eight APX genes have been reported, which includes two mitochondrial APXs (*OsAPX5* and *OsAPX6*), two chloroplastic APXs (*OsAPX7* and *OsAPX8*), two cytosolic APXs (*OsAPX1* and *OsAPX2*), and two peroxisomal APXs (*OsAPX3* and *OsAPX4*) (Teixeira et al. 2006). OsAPX1 and OsAPX2 were found to have significant role in stress tolerance in rice (Zhang et al. 2013).

Bonifacio et al. (2011) showed that the mutants of rice for cytosolic APXs displayed major changes in the redox status of the cell as indicated by changes in prooxidant/antioxidant ratio, thereby inducing ROS signaling networks and imparting stress tolerance. Similarly, there are evidences of evolution of large multigene families of some antioxidative enzymes like GST, TRX, and GRX with diverse functions that cope with a diversity of environmental stresses (Dixon and Edwards 2010). The members of these classes of multigene families are evolved as a key regulator in the maintenance of ROS homeostasis in crop plants. A thioredoxin gene *OsTRXh1*, which encodes h-type TRX in rice, is found to control the redox state of the apoplast and significantly take part in originating developmental and stress signals and responses (Zhang et al. 2013). The OsTRXh1 protein is found to have antioxidant activity and secreted into the apoplastic space. Therefore, the overexpression of *OsTRXh1* causes significant reduction in the accumulation of ROS, particularly H_2O_2, under salt stress, and reduces the expression of genes of salt-sensitive rice.

The ROS in signaling network needs a delicate balancing act between ROS formation and scavenging pathways for maintaining the endogenous titer of ROS below threshold and nontoxic levels for ensuring the subsequent downstream processes. Past studies showed that the formation of ROS and the activity of various antioxidant enzymes significantly changed under environmental stresses (Bhattacharjee 2005, 2012; Selote and Khanna-Chopra 2010; Turan and Ekmekci 2011). Presently, there are ample of evidences concerning the mechanisms by which antioxidant defense systems are modulated under the influence of abiotic stresses in crops for evoking acclamatory stress responses. Inherent with this antioxidative defense regulation is ROS production and associated redox signaling that integrated with the physiology of hormone and small molecules.

The growth retardants ABA, JA, SA, and ethylene are the key regulators of stress response in plants and regulate the expression of a large number of stress-responsive genes by a complex signaling network so as to confer stress tolerance. Growth retardant-induced stress tolerance is significantly linked with the activation and efficiency of antioxidant defense systems, which necessarily restores redox homeostasis and protects plant cells against nonspecific oxidative damages (Zhang et al. 2012a, 2014). Both dehydration stress and exogenous ABA application trigger the changes in the redox status of the plant tissue with enhanced formation of ROS, followed by initiation of redox signaling and the activation of the upregulation of antioxidant defense system in crops (Ye et al. 2011).

Plant recruits several important second messengers like Ca^{2+} and ROS in ABA-induced antioxidant defense mechanisms (Jiang and Zhang 2003; Hu et al. 2007). A protein kinase OsDMI3, which is essentially a Ca^{2+}/CaM-dependent one, is found to be associated with ABA-induced increase upregulation in the expression of two antioxidative defense genes and their corresponding activities (SOD and CAT) in rice. ABA-induced oxidative burst is also associated with activation of OsDMI3, which enhances H_2O_2 production by upregulating the expression of NADPH oxidases (Shi et al. 2012). When adjudged, OsDMI3 seems to work upstream of OsMPK1, to regulate the activities of antioxidant enzymes and the formation of H_2O_2 in rice under stress. Similarly, rice histidine kinase OsHK3 works upstream of OsDMI3 and OsMPK1 and is associated with abiotic stress-induced ABA-mediated antioxidant defense mechanism (Wen et al. 2015). Zhang et al. (2012a, b, 2014) described that ZFPs like ZFP36 and ZFP182 are involved in ABA-induced upregulation of antioxidant defense. It was also found that both the ABA-induced oxidative burst and activation of OsMPKs upregulate the expression of *ZFP36* and *ZFP36* and subsequently promote the expression of genes like NADPH oxidase and MAPK responsible for biphasic generation of ROS H_2O_2 in ABA signaling (Zhang et al. 2014). Both ABA and H_2O_2, in maize, significantly increased the expression and the activity of ZmMPK5, as a part of stress acclamatory defense mechanism. Upregulation of ZmMPK5, which in turn, increases the expression and the activity of NADPH oxidase, is also responsible for ROS generation. So, there exists a positive feedback loop involving ZmMPK5, NADPH oxidase, and ROS, in ABA signaling (Zhang et al. 2006, 2007; Hu et al. 2007; Lin et al. 2009). In 2007, Hu et al. described that calcium signature through Ca^{2+}/CaM pathway is required for triggering ABA-induced antioxidant defense mechanism, which might function both upstream and downstream of ROS production in plants. In this ABA-dependent Ca^{2+}/CaM pathway, the protein kinase ZmCCaMK was reported to be essential for antioxidant defense, where ROS-induced NO production is involved additionally in the activation of protein kinase ZmCCaMK (Ma et al. 2012).

Growth inhibitors brassinosteroids (BR) are basically steroidal hormones having profound influence on plant responses to biotic and abiotic stresses (Bajguz and Hayat 2009; Divi and Krishna 2009; Yang et al. 2011; Zhu et al. 2013a). Several studies have shown that BR can upregulate antioxidant defense mechanism for conferring stress tolerance in crops. Zhang et al. (2013) reported that the BR-induced antioxidant defense operates via ZmMPK5 and NADPH oxidase for the generation of ROS signaling in maize. A BR-dependent phosphorylation of ZmMPK5 is required for the activation of 65 KDa microtubule-associated protein (MAP65), ZmMAP65-1a, as a part of antioxidative defense mechanism in plant (Zhu et al. 2013b). Lately, it is also found that the BR-induced antioxidant defense in maize needs both Ca^{2+} signature and activation of CCaMK gene, *ZmCCaMK* (Yan et al. 2015).

An upregulation in the first line of antioxidative defense involving SOD activity in response to water deficit stress was noticed in different cultivars of rice and mung bean (Zlatev et al. 2006; Sharma and Dubey 2005a, b). The activity of the same enzyme was also found to be enhanced significantly under drought in the leaves of

Trifolium repens L. (Chang-Quan and Rui-Chang 2008). Salt stress also caused enhancement in SOD activity in other crops like chickpea (Kukreja et al. 2005) and tomato (Gapińska et al. 2008). An isozymic variation of SOD is also noticed, and the three isozymic forms of SOD have been found to be expressed in chickpea under NaCl stress (Eyidogan and Öz 2007). Salt stress tolerance for the transgenic *Arabidopsis* was found to be associated with overexpression of Mn-SOD (Wang et al. 2004). Interestingly, the SOD activity was found to be augmented by UV-B radiation in *Arabidopsis*, rice, pea, and wheat but remains unchanged in barley and soybean. Supplementation of UV-B increased SOD activity in wheat and mung bean but exhibited differential responses among soybean cultivars (Agrawal et al. 2009).

The activity of CAT, which renders H_2O_2 detoxification, was found to enhance under stress, especially in drought-sensitive varieties of wheat (Simova-Stoilova et al. 2010). Salt stress upregulates the activity of CAT in roots and leaves (Eyidogan and Öz 2007; Kukreja et al. 2005). Heavy metal stress was also found to augment the activity of CAT in crops like *Phaseolus aureus*, *Pisum sativum*, barley, and sunflower (Sreedevi and Krishnan 2012). When enzymatic antioxidant profiling of different accessions of wheat with differential drought-responsiveness was compared, it was observed that the drought tolerance of the genotypes has direct relationship with higher antioxidant enzyme activity like APX and CAT and AA content and lower H_2O_2 and MDA content (Sairam et al. 1998).

Chakraborty and Bhattacharjee (2015) in a study showed that the differential efficacy of ASC-GSH cycle for the processing of hydrogen peroxide between the salt-tolerant (SR26B) and salt-sensitive (Ratna) rice cultivar under chilling, heat, and oxidative stress of higher magnitude contributes significantly in conferring differential redox-regulatory properties in the germinating tissue of two experimental rice cultivars. The role of redox homeostasis in which ROS (ascorbate-glutathione) system acts at metabolic interface is not only for preventing oxidative damages but also for the origin of endogenous redox cues necessary for upregulation of expression of antioxidative genes and redox-regulatory proteomes.

The salt-tolerant cultivar SR26B resisted from chilling, heat, and oxidative stress of higher magnitude due to its preparedness to maintain redox homeostasis by upregulation of gene expression of antioxidative enzymes (particularly of ascorbate-glutathione cycle) and hence the better capacity of redox regulation and mitigation of oxidative damage to the germinating tissue. The transcript abundance of genes of *SodCc2*, *CatA*, *OsAPx2*, and *GRase* coding, respectively, for cytosolic Cu/Zn superoxide dismutase (Cu/ZnSOD), catalase (CAT), ascorbate peroxidase (APOX), and glutathione reductase (GR) exhibited differential relative expression for salt-tolerant and susceptible cultivars of rice under extremes of temperature (Chakraborty and Bhattacharjee 2015).

Subcellular localization of APX exhibited overexpression of the enzyme in the chloroplasts of *Nicotiana tabacum*, necessary for the reduction of H_2O_2 associated with drought tolerance (Badawi et al. 2004). In a study with a salt-tolerant and salt-sensitive rice cultivar, RoyChoudhury et al. (2012c) showed the differential activities of GPX and APX in IR-29 (salt-sensitive) and Nona Bokra (salt-tolerant) rice cultivars under heavy metal stress, with more enhanced activity in Nona Bokra, hinting at the cross tolerance of the rice cultivar. The POX and CAT activity is also found to be strongly associated with rising heavy metal concentration of growing media of *Vigna radiata* (Roychoudhury and Ghosh 2013). *Vaccinium myrtillus L.*, an acid-tolerant and heavy metal-resistant species, was found to accumulate significantly higher amount of GSH, nonprotein thiols, and proline, and activity of GPX was elevated (Kandziora-Ciupa et al. 2013). Augmentation in MDHAR and DHAR activities in tobacco and *Arabidopsis* (Eltayeb et al. 2007; Ushimaru et al. 2006) resulted in improved salt tolerance. Abiotically stressed rice seedlings displayed increased activities of ascorbate-glutathione pathway enzymes MDHAR, DHAR, and GR, for better processing and detoxification of H_2O_2 (Sharma and Dubey 2005a, b).

Transgenic tobacco overexpressing a tocopherol biosynthesis enzyme *Arabidopsis VTE1* showed reduced oxidative membrane damage (assessed in terms of MLPO, electrolyte leakage, and H_2O_2 content) (Liu et al. 2008). The mutants of *Arabidopsis vte1* and *vte4*, lacking in their ability to synthesize vitamin E and failing to maintain cellular Na^+/K^+ ion balance became sensitive to salt stress, as can be vouched by their reduced growth and oxidative damages (Ellouzi et al. 2013). Similarly exposure to UV-B leads to diminish in α-tocopherol titer in plants, hinting at reactions with toxic secondary oxidative products like lipid radicals (Jain and Kataria 2003).

The role of antioxidant pigment carotenoids under stress can be vouched from the evidences of significantly greater accumulation of carotenoid molecules per chlorophyll unit under drought stress with their ability to prevent photooxidative damages (Munné-Bosch and Alegre 2000). PEG 6000 induced dehydration stress to rice seedlings led to enhancement in accumulation of flavonoids and phenolics to the tolerant rice cultivar Pokkali, as compared to the sensitive one like IR-29 (Basu et al. 2010a). Solar UV-B radiation caused oxidative stress in two isolines of soybean cv. Clark and altered redox status. The tolerant line suffers lesser magnitude of oxidative stress by reducing SOD activity and enhancing the activities of APX, CAT, and GR. This resulted in better processing of H_2O_2 through ascorbate-glutathione pathway. In fact, the susceptible line had greater oxidative stress than the normal line, in spite of its ascorbate-glutathione pathway, as compared to the tolerant normal line, even under UV-B exclusion. It indicates that the enhanced sensitivity in the susceptible line, under UV-B exclusion, was due to the absence of flavonoids in the epidermal cells, necessary for ROS scavenging (Xu et al. 2008; Collins 2001).

The amino acid proline is a bona fide cellular osmolyte, primarily implicated in stress response as an osmoprotectant and an alternative sink for energy for regulation of redox potential of cell. Proline is found to increase significantly in drought-tolerant cultivars than sensitive cultivars of chickpea particularly under drought stress (Mafakheri et al. 2010). Rice seedlings, exposed to NaCl stress, showed an upregulation in the synthesis of anthocyanin and proline. The salt-tolerant cultivars Nona Bokra, SR26B, etc. showed to accumulate significantly higher amount of proline as compared to sensitive cultivars (Chakraborty and Bhattacharjee 2015; Roychoudhury et al. 2008). These tolerant cultivars exhibited better redox-regulatory properties as compared to their counterpart and survive under stress. The content of these nonenzymatic antioxidants like flavonoids and proline was found to be significantly higher in both salinity-stressed and control conditions in salt-tolerant cultivars as compared to its counterpart, i.e., salt-sensitive one, as evident by their reduced accumulation of oxidative stress biomarkers (Bhattacharjee and Dey 2017; Chutipaijit et al. 2009).

2.4 ROS-Antioxidant Interaction at Metabolic Interface Determines Redox Signaling Under Environmental Stress

Loss of redox homeostasis due to overproduction of reactive oxygen species (ROS) or incompetence of antioxidative defense system in plants creates oxidative stress, causing damage to all major cellular macromolecules (Gill and Tuteja 2010; Bhattacharjee 2005; Wang et al. 2014; RoyChoudhury and Basu 2012). Abiotic stresses, particularly drought, salinity, and extremes of temperature experienced by crops, often instigate oxidative stress. These stress responses, particularly drought and salinity, in different plants depend on the intrinsic strategy of plant and the duration and severity of stress, which is essentially an outcome of ROS-antioxidant interaction at metabolic interface (Fig. 2.1). An extended period of dehydration stress results in oxidative damage due to the excess generation of ROS (Cruz and Helena 2008; Zhu 2000, 2002; Ingram and Bartels 1996). In optimum environmental condition, both the generation and detoxification of ROS generally maintain a balance (Foyer and Noctor 2000; Porcel et al. 2003). In fact, in optimized state, ROS is not

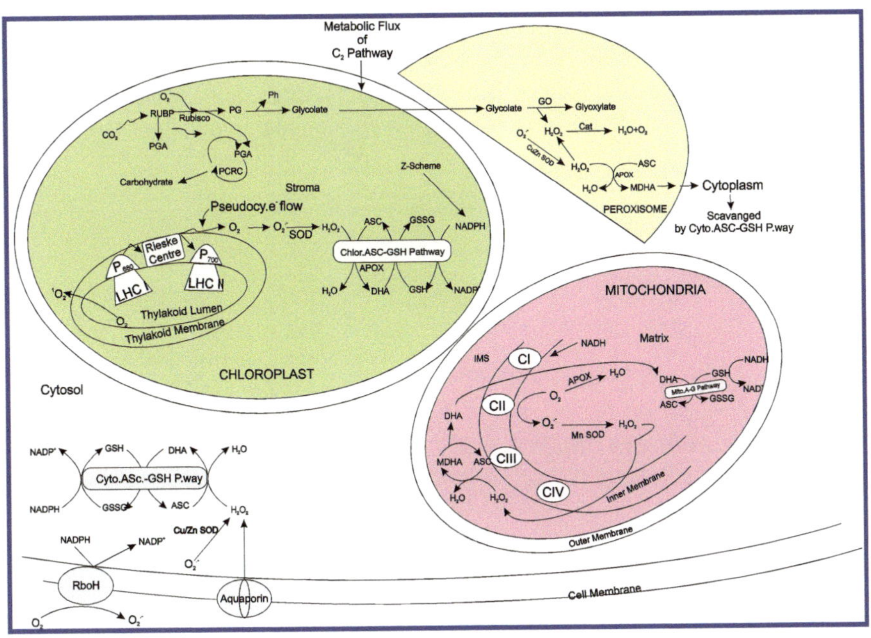

Fig. 2.1 ROS-antioxidant interaction involved in the maintenance of redox homeostasis in plant cell. (Detail in text)

instigating its unwanted lethal oxidative role, rather initiates signaling processes and eventually causing expression of a number of stress acclamatory and developmental genes useful in numerous physiological plant processes like stress acclimation, adventitious root formation, guard cell movement, apoptosis, etc. But in situations where the accumulation of prooxidants exceeds the amount that can be removed by antioxidative defense system, both enzymatic (SOD,CAT,APX, GR, MDHAR, DHAR, GPX, GOPX, GST, etc.) and nonenzymatic (ascorbic acid, glutathione, polyphenolic compounds, proline, anthocyanin, carotenoids, betacyanin, alkaloids, nonprotein amino acids, and α-tocopherols) antioxidant systems (Table 2.1), unwanted oxidative damage to cellular component takes place (Gill and Tuteja 2010, Bhattacharjee 2005; Vaidyanathan et al. 2003; Hasanuzzaman et al. 2013; Golldack et al. 2014; Gaber 2010). In this context, the differential antioxidant capacities, both enzymatic and nonenzymatic, were found to be linked with modification of competence of redox-regulatory metabolisms of closely related germplasms (Table 2.2).

To combat drought stress, loss of water by closure of stomata of plants is very natural. But this process of maintenance of water-use economy also reduces the CO_2 uptake by the plant and photosynthetic carbon reduction cycle, causing disturbances in redox homeostasis and loss of well-tuned balance between ROS production and scavenging processes (Mittler 2002; Chen et al. 2013; Basu et al. 2010b; Shinozaki and Yamaguchi-Shinozaki 2007). Essentially, all ROS like $^1O_2$, $OH^{\cdot}$, $O_2^{\cdot-}$, H_2O_2, $RCO^{\cdot}$, $RO_2^{\cdot}$ etc. can instigate oxidative damage under drought stress through protein oxidation, membrane lipid peroxidation, enzyme inhibition, and nucleic acid damage (Grene 2002). Genotype-specific variations of antioxidative defense mechanisms are noticed under drought stress. From various previous research, it is apparent that antioxidative enzymes exhibited a general upregulation in their activity to confer stress tolerance (Sun et al. 2010; Cruz and Helena 2008; Foyer and Noctor 2005a, b). For example, SOD activity was found to reduce significantly under drought to keep stomatal conductance marginally

Table 2.1 Upregulation of antioxidant defense in response to abiotic stress

Stresses	Antioxidant enzymes upregulated	Plant species	References
Drought	SOD, GPX, APX, MDHAR, DHAR, and GR OsMTla	*Oryza sativa*	Bhattacharjee (2012), Sharma and Dubey (2005); Sharma et al. (2012) and Yang et al. (2009)
	SOD, CAT, and GPX	*Beta vulgaris*	Sayfzadeh and Rashidi (2011) and Sharma et al. (2012)
	SOD, APX, and GR	*Triticum sativum*	Sgherriei et al. (2000) and Sharma et al. (2012)
Salinity	SOD, CAT, GPX, APX, GR	*Oryza sativa*	Mishra et al.; Sharma et al. (2012)
	GPX	*Oryza sativa*	Mittal and Dubey (1991)
	APX, GST, GLX I, SAM synthase	*Hordeum vulgare*	Witzelei et al. (2009)
Temperature stress	Mn-SOD, DHAR, GR	*Oryza sativa*	Sato et al. (2011) and Takesawa et al. (2002)

Table 2.2 Antioxidant competence and differential stress resistance attributes of crop varieties

Stress-resistant cultivar	Stress susceptible cultivar	Differential antioxidant competence	References
Oryza sativa L, cultivar SR26B	*Oryza sativa* L, cultivar Ratna	Better antioxidative competence in Cv. SR26B compared to Ratna in terms of activities of antioxidative defense enzymes (CAT, SOD, APOX, GR), ROS accumulation, oxidative damages to membrane components	Bhattacharjee (2012) and Chakraborty and Bhattacharjee (2015)
Oryza sativa L, cultivar SR26B	*Oryza sativa* L, cultivar Ratna	Increased transcript abundance of antioxidative defense enzymes (CAT, SOD, APOX, GR),	Chakraborty and Bhattacharjee (2015)
Hordeum marinum	*Hordeum vulgare*	Increased SOD, CAT activity in the halophyte	Seckin et al. (2010)
Lycopersicon pennellii	*Lycopersicon esculentum*	Increased SOD activities in chloroplast, mitochondria, and peroxisomes of the halophyte	Shalata et al. (2001)
Lycopersicon pennellii	*Lycopersicon esculentum*	Constitutively higher peroxisome CAT activity and further increased during salt stress in the halophyte	Mittova et al. (2000, 2003) and Shalata et al. (2001)
Lycopersicon pennellii	*Lycopersicon esculentum*	Constitutively higher APX activity and further increased during salt stress in chloroplast, peroxisome, and mitochondrial fraction of the halophyte	Mittova et al. (2000, 2003) and Shalata et al. (2001)
Hordeum marinum	*Hordeum vulgare*	Increased POX activity in the halophyte	Seckin et al. (2010)
Lycopersicon pennellii	*Lycopersicon esculentum*	MDAR activity increased in the halophyte but decreased in the glycophyte	Mittova et al. (2000, 2003) and Shalata and Neumann (2001)
Hordeum marinum	*Hordeum vulgare*	Activities of isoenzymes GR1, 3, 6, 7 increased in halophyte	Seckin et al. (2010)
Lycopersicon pennellii	*Lycopersicon esculentum*	Constitutively higher GR activity in mitochondrial and peroxisome fractions of the halophyte but decreased during stress in both the glycophyte and halophyte	Mittova et al. (2003)

more and avoid complete inhibition of PCRC. This phenomenon has wide implication in adaptive stress response of plants, as the event significantly reduces the chances of redox imbalance and further episodes of oxidative stresses (Cruz and Helena 2008). It is also found that the defense enzymes like SOD, CAT, APX, DHAR, MDHAR, GR, and GPX/GST were increased in their activity either alone or in a concerted manner during drought stress. Nonenzymatic antioxidants or ROS quenchers like ascorbic acid, glutathione, phenolics, flavonoids, carotenoid, anthocyanin, betacyanin, etc. also play a vital role to restore the redox homeostasis by buffering the endogenous concentrations of ROS in plant cell either by quenching the ROS directly or indirectly by reducing their probability of formation (Shao et al. 2008; Dat et al. 2013).

That these ROS-antioxidant interactions at metabolic interface are buffering the endogenous level of ROS and initiate subsequent redox signaling could be vouched by the upregulation in activities of several signaling intermediates. For example, the plasma membrane-bound protein kinases or receptor-like kinases (RLKs) act as signal transducers under drought stress- induced oxidative stress (Xiong et al. 2002; Marshall et al. 2012). These RLKs are specific kinases with transmembrane domain and an intracellular domain with kinase activity or without having any extracellular domain or simply a intracellular kinase domain (Shiu and Bleecker 2001, 2003; Jurca et al. 2008; Vij et al. 2008). RLKs sense change in redox status of the cell under stress causing either homo- or heterodimerization, subsequently followed by autophosphorylation of the cytoplasmic kinase domain and subsequently transphosphorylation of other moieties (Morris and Walker 2003). GROWTH UNDER DROUGHT KINASE (GUDK), an RLK gene of rice, is found to be highly drought inducible and redox regulatory and required for combating stress for optimum grain yield. Molecular genetic experiment, with mutants having loss of function of GUDK, showed a reduction in photosynthetic efficiency under drought, salinity, and osmotic stress both at vegetative stage and seedling stage. These experiments and transactivation assays confirmed that GUDK is required for activation of stress genes by transcription factor OsAP37 and are associated with grain yield of rice (Ramegowda et al. 2014). For several stress-related redox-sensitive genes, miRNAs function as significant posttranscriptional regulators for modulating their expression. These modulator miRNAs combined with their target genes constitute large redox-regulatory networks which subsequently control antioxidant defense, stress responses, growth inhibitor response, osmoprotection, etc. (Ding et al. 2013a, b).

Salinity stress can prevent water uptake which induces secondary water deficit stress similar to drought stress. So a similar pattern of stress response and signaling between salinity and drought stress was observed (Tippmann et al. 2006; Aghaei et al. 2009). A noteworthy feature of signaling cross talk between these two stresses (drought and salinity) was also observed (Shinozaki et al. 2003; Ashraf and Akram 2009; Athar et al. 2008). Under both drought and salinity, stress is sensed or recognized by ROS and modulation of intracellular calcium (Ca^{2+}), either directly or mediated by separate receptors. There exist both feed-forward and backward interactions between ROS and Ca^{2+}. The redox signals which work through the specific phosphorylation cascades (MAPK) and or calcium-dependent protein kinases (CDPKs) interact with specific transcription factors (TF) and redox-sensitive genes and under stress (Fig. 2.2). The origin of redox signal, which largely depends upon antioxidant makeup and their efficiency, has direct impact on gene expression under stress. Reduction in PCRC in drought and salinity stress causes the change in the redox status of both chloroplast and mitochondria as the events promote ROS production. Now the detox-scavenging efficiency of antioxidative event will determine the internal redox cue and the downstream signaling pathway (Tippmann et al. 2006). In fact, for the dual fate of ROS, antioxidant systems play supplementary or additional role in redox signaling for optimizing the amount of ROS in contrary to

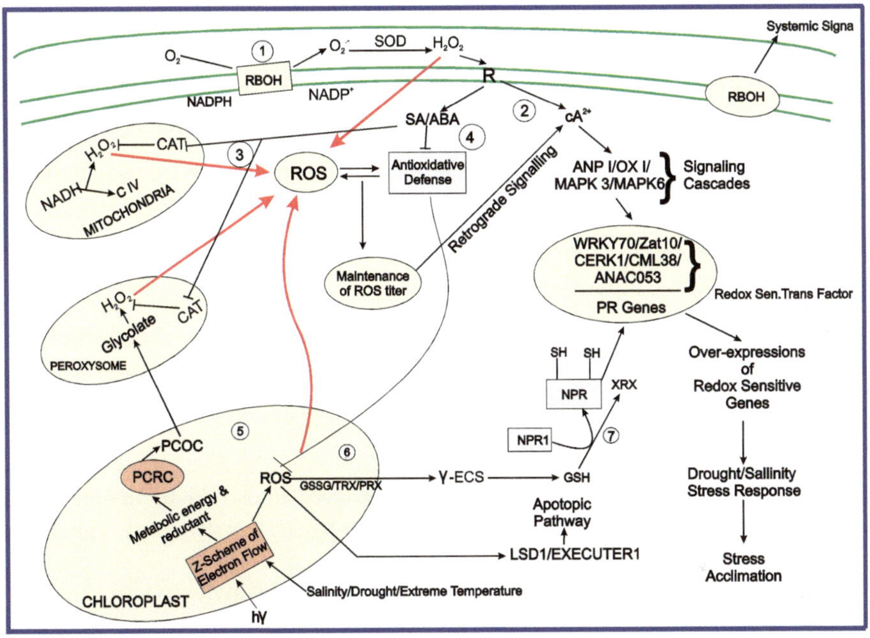

Fig. 2.2 Conceptual scheme showing interaction of redox circuits, with other signaling cascades under environmental stress. (Detail in text)

the entire removal of ROS. Therefore, the objective of enhanced production of ROS during stress may not be visualized as a toxic consequence of the cell, rather with the help of antioxidant buffering, they can act as signal to initiate the acclimatory stress response. It has been deciphered that these redox signaling pathways exploit MAPK cascades, Ca^{2+} signature, histidine kinase sensors, transcription factors, etc. (Apel and Hirt 2004).

2.5 Antioxidant and Redox Sensing Under Environmental Stress

It is now well established that ROS-mediated signaling or redox signaling in plants under environmental stress largely involves heterotrimeric G-proteins (Joo et al. 2005), Ca^{2+} signature, protein phosphorylation cascades involving specific MAPKs, and protein Tyr phosphatases (Kovtun et al. 2000; Gupta and Luan 2003; Rentel et al. 2004). Though the exact mechanism or the molecular basis of kinase pathway activation by ROS signaling remains to be established, the redox sensor thiol group oxidation seems to play a significant role in this aspect. The well-characterized redox signal transduction system, i.e., the stromal ferredoxin-thioredoxin system, which functions in the regulation of PCRC, is an ideal example for the same (Fig. 2.1). Here the signal transmission takes place through disulfide-thiol conversion or exchange mechanism in target enzymes which is found to take place by light-induced reduction in the thioredoxin redox potential (Setterdahl et al. 2003). Thiol groups are also extremely important in other types of redox signal transduction like ETR1 receptor kinase-based ROS sensing (Desikan et al. 2005). This thiol-based redox regulation and signaling may be significant in plant acclimation to unfavorable environmental conditions, particularly in a situation where redox interactions play a key role in the orchestration of stress response. The oxidation of thiol containing ROS gives relatively stable oxidation products with changed architectural and biochemical properties (Bauer et al. 1999; Delauney et al. 2002). Apart from this, other oxidized species of cysteine sulfur (sulfenic acid, GSH-Cys, sulfenyl amides, and sulfur-metal bonds) might be vital in redox sensing mechanisms. Essentially all these signaling mechanisms are the outcome of direct effect of ROS on -SH containing transcription factors or proteins of signal transduction pathways operating under stress. Examples of overexpression of transcription factors under redox signaling are JERF3, Zat 10, and Zat12, which subsequently modulates the expression of redox-sensitive genes necessary for affording tolerance to stress (Wu et al. 2008; Miller et al. 2010). Furthermore overexpression MAPKK1 in *Arabidopsis* is also taking place by ROS-antioxidant signaling network under stress. The overall result of ROS-antioxidant interaction at metabolic interface is that deficiency in MKK1 resulted in oxidative stress and enhanced stress sensitivity (Xing et al. 2008).

Several works in the last decade strongly convey the role of antioxidative enzymes apart from their traditional ROS scavenging role (Miller et al. 2007). It is now gradually apparent that antioxidants have putative role in signaling (Fig. 2.2). As an evidence, knockout of APX1 gene is shown to perform better than wild plants under stress (Miller et al. 2007). Similarly, *Arabidopsis* mutants deficient in APX1 cause induction of several antioxidative defense genes under stress (Davletova et al. 2005). Similarly, the antisense CAT1 and APX1 in tobacco are found to prone toward oxidative stress-induced damages, but their corresponding double antisense lines became significantly tolerant to oxidative stress (Rizhsky et al. 2002). The overexpression of Zat 7 associated with knockout *Apx* plant also causes enhanced expression of other defense transcripts such as AOX1, WRKY 70, etc., conferring significant role in stress tolerance (Ciftci-Yilmaz et al. 2007). The parallel function of antioxidants and redox sensing mechanisms, apart from its conventional role in conferring stress tolerance of plants, can be further substantiated by biochemical and structural elucidation of ROS-responsive transcript- encoding TFs (transcription factors) and other intermediate signaling proteins (Fig. 2.2).

2.6 Conclusion

Evolution has equipped plants with wider options of defense measures in the form of antioxidants. This chapter provides an insight into the fact how the two arms of the antioxidant machinery, i.e., enzymatic and nonenzymatic antioxidants, work individually or in tandem to mitigate the damaging effects of ROS and develop detox-scavenging mechanisms necessary for conferring tolerance against environmental odds. This chapter also provides an interpretation of how plant cell through a series of interacting redox components with different mechanisms of antioxidant buffering capacities determines the fate of the cell, particularly under unfavorable environmental conditions. In fact, the environmental cues trigger specific ROS-antioxidant interaction, initiating the formation of endogenous redox cues, being sensed by redox sensors for inducing ROS signaling. It is the antioxidants which through its interaction with its counterpart ROS that determine the fate of oxidative stress. So in these broad redox signaling networks, antioxidants are not only the ROS scavengers or passive bystander but rather invariably function as key signaling components which establish a flexible and dynamic metabolic frontier between perception of environmental cues and ultimate physiological response.

Although the present text dealt with the highly compartmentalized nature of different types of antioxidants and their effect in restoring redox homeostasis, the regulation of the delicate balance between generation and scavenging of ROS, in this regard, needs to be better explored. In this regard, though the origin of redox signal where antioxidants and ROS interacting at metabolic interface is well known, several interesting issues in this aspect remain uncultivated, like the interaction between ROS and other signaling mechanisms in regulation of the synthesis and action of redox buffers or antioxidant, particularly under environmental stresses and developmental processes. In the future, advanced technologies like Ca^{2+} and ROS imaging, mutant-based study of hormonal signal transduction, identification of phosphorylation cascades, advanced functional genomics coupled with proteomics and metabolomics, etc. can lead to better understanding of ROS-antioxidant interaction at metabolic interface and adjudging the proper position of antioxidants in redox signaling hub.

References

Agati G, Azzarello E, Pollastri S, Tattini M (2012) Flavonoids as antioxidants in plants: location and functional significance. Plant Sci 196:67–76

Aghaei K, Ehsanpour AA, Komatsu S (2009) Potato responds to salt stress by increased activity of antioxidant enzymes. J Integr Plant Biol 51:1095–1103

Agrawal SB, Singh S, Agrawal M (2009) Ultraviolet-B induced changes in gene expression and antioxidants in plants. Adv Bot Res 52:47–86

Apel K, Hirt H (2004) Reactive oxygen species: Metabolism, oxidative stress, and signal transduction. Annu Rev Plant Biol 55:373–399

Asada K (1999) The water-water cycle in chloroplasts: scavenging of active oxygens and dissipation of excess photons. Annu Rev Plant Physiol Plant Mol Biol 50:601–639

Asada K (2000) The water-water cycle as alternative photon and electron sinks. Phil Trans R Soc Lond B Biol Sci 355:1419–1431

Asada K (2006) Production and scavenging of reactive oxygen species in chloroplasts and their functions. Plant Physiol 141:391–396

Ashraf M, Akram NA (2009) Improving salinity tolerance of plants through conventional breeding and genetic engineering: analytical comparison. Biotechnology 27:744–752

Athar HUR, Khan A, Ashraf M (2008) Exogenously applied ascorbic acid alleviates salt-induced oxidative stress in wheat. Environ Exp Bot 63:224–223

Badawi GH, Yamauchia Y, Shimada E, Sasaki R, Kawano N (2004) Enhanced tolerance to salt stress and water deficit by overexpressing superoxide dismutase in tobacco (*Nicotiana tabacum*) chloroplasts. Plant Sci 66:919–928

Bajguz A, Hayat S (2009) Effects of brassinosteroids on the plant responses to environmental stresses. Plant Physiol Biochem 47:1–8

Barnes J, Zheng Y, Lyons T (2002) Plant resistance to ozone: the role of ascorbate. In: Omasa K, Saji H, Youssefian S, Kondo N (eds) Air pollution and plant biotechnology—prospects for phytomonitoring and phytoremediation. Springer, Tokyo, pp 235–252

Basu S, Roychoudhury A, Saha PP, Sengupta DN (2010a) Differential antioxidative responses of indica rice cultivars to drought stress. Plant Growth Regul 60:51–59

Basu S, Roychoudhury A, Saha PP, Sengupta DN (2010b) Comparative analysis of some biochemical responses of three indica rice varieties during polyethyleneglycol-mediated water stress exhibits distinct varietal differences. Acta Physiol Plant 32:551–563

Bauer CE, Elsen S, Bird TH (1999) Mechanisms for redox control of gene expression. Annu Rev Microbiol 53:495–523

Bhattacharjee S (2005) Reactive oxygen species and oxidative burst: roles in stress, senescence and signal. Curr Sci 89:1113–1121

Bhattacharjee S (2012) The language of reactive oxygen species signaling in plants. J Bot 2012:22

Bhattacharjee S, Dey N (2017) Redox metabolic and molecular parameters for screening drought-tolerant indigenous aromatic rice cultivars. Physiol Mol Biol Plants 24(1):7–23

Biehler K, Fock H (1996) Evidence for the contribution of the Mehler-peroxidase reaction in dissipating excess electrons in drought stressed wheat. Plant Physiol 112:265–272

Boguszewska D, Grudkowska M, Zagdańska B (2010) Drought responsive antioxidant enzymes in potato (*Solanum tuberosum* L.). Potato Res 53:373–382

Bonifacio A, Martins MO, Ribeiro CW, Fontenele AV, Carvalho FE, Margis-Pinheiro M, Silveira JA (2011) Role of peroxidases in the compensation of cytosolic ascorbate peroxidase knockdown in rice plants under abiotic stress. Plant Cell Environ 34:1705–1722

Brown JE, Khodr H, Hider RC, Rice-Evans CA (1998) Structural dependence of flavonoid interactions with Cu2+ ions: implications for their antioxidant properties. Biochem J 330(3):1173–1178

Cao G, Sofic E, Prior RL (1997) Antioxidant and prooxidant behavior of flavonoids: structure-activity relationships. Free Radic Biol Med 22(5):749–760

Chakraborty A, Bhattacharjee S (2015) Differential competence of redox-regulatory mechanism under extremes of temperature determines growth performances and cross tolerance in two indica rice cultivars. J Plant Physiol 176:65–77

ChalapathiRao ASV, Reddy AR (2008) Glutathione reductase: a putative redox regulatory system in plant cells. In: Khan NA, Singh S, Umar S (eds) Sulfur assimilation and abiotic stress in plants. Springer, Berlin, pp 111–147

Chang-Quan W, Rui-Chang L (2008) Enhancement of superoxide dismutase activity in the leaves of white clover (*Trifolium repens* L.) in response to polyethyleneglycol-induced water stress. Acta Physiol Plant 30:841–847

Chen Z, Gallie DR (2005) Increasing tolerance to ozone by elevating folia ascorbic acid confers greater protection against ozone than increasing avoidance. Plant Physiol 138:1673–1689

Chen Z, Gallie DR (2006) Dehydroascorbate reductase affects leaf growth, development and function. Plant Physiol 142:775–787

Chen Q, Zhang M, Shen S (2010) Effect of salt on malondialdehyde and antioxidant enzymes in seedling roots of Jerusalem artichoke (*Helianthus tuberosus* L.). Acta Physiol Plant 33:273–278

Chen Q, Tao S, Bi X, Xu X, Wang L, Li X (2013) Research progress in physiological and molecular biology mechanism of drought resistance in rice. Am J Mol Biol 3(2):102–107

Chutipaijit S, Cha-Um S, Sompornpailin K (2009) Differential accumulations of proline and flavonoids in indica rice varieties against salinity. Pak J Bot 41:2497–2506

Ciftci-Yilmaz S, Morsy MR, Song L, Coutu A, Krizek BA, Lewis MW, Warren D, Cushman J, Connolly EL, Mittler R (2007) The ear-motif of the C2H2 zinc-finger protein ZAT7 plays a key role in the defense response of *Arabidopsis* to salinity stress. J Biol Chem 282:9260–9268

Collins A (2001) Carotenoids and genomic stability. Mutat Res 475:1–28

Creissen GP, Broadbent P, Kular B, Reynolds H, Wellburn AR, Mullineaux PM (1994) Manipulation of glutathione reductase in transgenic plants: implications for plant responses to environmental stress. Proc R Soc 102B:167–175

Cruz CD, Helena M (2008) Drought stress and reactive oxygen species Production, scavenging and signalling. Plant Signaling Behavior 3(3):156–165

Das K, Roychoudhuri A (2014) Reactive oxygen species (ROS) and response of antioxidants as ROS-scavengers during environmental stress in plants. Front in Env Sci 2. https://doi.org/10.3389/fenvs.2014.00053

Dat JF, Pellinen R, Beeckman T, Van De Cotte B, Langebartels C, Kangasjarvi J, Inze D, Van Breusegem F (2013) Changes in hydrogen peroxide homeostasis trigger an active cell death process in tobacco. Plant J 33:621–632

Davletova S, Rizhsky L, Liang H, Shengqiang Z, Oliver DJ, Coutu J, Shulaev V, Schlauch K, Mittler R (2005) Cytosolic ascorbate peroxidase 1 is a central component of the reactive oxygen gene network of *Arabidopsis*. Plant Cell 17:268–281

del Río LA, Corpas FJ, Sandalio LM, Palma JM, Gómez M, Barroso JB (2002a) Reactive oxygen species, antioxidant systems and nitric oxide in peroxisomes. J Exp Bot 53:1255–1272

del Río LA, Sandalio LM, Palma JM, Corpas FJ, LópezHuertas E, Romero-Puertas MC, McCarthy I (2002b) Peroxisomes, reactive oxygen metabolism, and stress-related enzyme activities. In: Baker A, Graham I (eds) Plant peroxisomes biochemistry, cell biology and biotechnological applications. Kluwer Academic Publishers, Dordrecht, pp 221–258

Delauney A, Pflieger D, Barrault MB, Vinh J, Toledano MB (2002) A thiol peroxidase is an H_2O_2 receptor and redox transducer in gene activation. Cell 111:1–11

Desikan R, Hancock JT, Bright J, Harrison J, Weir I, Hooley R, Neill SJ (2005) A role for ETR1 in hydrogen peroxide signaling in stomatal guard cells. Plant Physiol 137:831–834

Ding Y, Cao J, Ni L, Zhu Y, Zhang A, Tan M et al (2013a) ZmCPK11 is involved in abscisic acid-induced antioxidant defence and functions upstream of ZmMPK5 in abscisic acid signalling in maize. J Exp Bot 64:871–884

Ding Y, Tao Y, Zhu C (2013b) Emerging roles of microRNAs in the mediation of drought stress response in plants. J Exp Bot 64(11):3077–3086

Divi UK, Krishna P (2009) Brassinosteroid: a biotechnological target for enhancing crop yield and stress tolerance. New Biotechnol 26:131–136

Dixon DP, Edwards R (2010) Glutathione transferases. Arabidopsis Book 8:e0131. https://doi.org/10.1199/tab.0131

Edwards EA, Rawsthorne S, Mullineaux PM (1990) Subcellular distribution of multiple forms of glutathione reductase in leaves of pea (*Pisum sativum* L.). Planta 180:278–284

Ellouzi H, Hamed K, Cela J, Müller M, Abdelly C, Munné-Bosch S (2013) Increased sensitivity to salt stress in tocopherol-deficient *Arabidopsis* mutants growing in a hydroponic system. Plant Signal Behav 8:e23136

Eltayeb AE, Kawano N, Badawi GH, Kaminaka H, Sanekata T, Morishima I, Shibahara T, Inanaga S, Tanaka K (2006) Enhanced tolerance to ozone and drought stresses in transgenic tobacco overexpressing dehydroascorbate reductase in cytosol. Physiol Plant 127:57–65

Eltayeb AE, Kawano N, Badawi GH, Kaminaka H, Sanekata T, Shibahara T, Inanaga S, Tanaka K (2007) Overexpression of monodehydroascorbate reductase in transgenic tobacco confers enhanced tolerance to ozone, salt and polyethylene glycol stresses. Planta 225:1255–1264

Eyidogan F, Oz MT (2005) Effect of salinity on antioxidant responses of chickpea seedlings. Acta Physiol Plant 29:485–493

Eyidogan F, Öz MT (2007) Effect of salinity on antioxidant responses of chickpea seedlings. Acta Physiol Plant 29:485–493

Fini A, Brunetti C, Di Ferdinando M, Ferrini F, Tattini M (2011) Stress- induced flavonoid biosynthesis and the antioxidant machinery of plants. Plant Signal Behav 6:709–711

Foyer CH, Noctor G (2000) Oxygen processing in photosynthesis: regulation and signaling. New Phytol 146(3):359–388

Foyer CH, Noctor G (2005a) Oxidant and antioxidant signaling in plants: a re-evaluation of the concept of oxidative stress in a physiological context. Plant Cell Environ 28:1056–1071

Foyer CH, Noctor G (2005b) Redox homeostasis and antioxidant signaling: a metabolic interface between stress perception and physiological responses. Plant Cell 17:1866–1875

Foyer CH, Shigeoka S (2011) Understanding oxidative stress and antioxidant functions to enhance photosynthesis. Plant Physiol 155(1):93–100

Gaber AM (2010) Antioxidative defense under salt stress. Plant Signaling Behavior 5(4):369–374

Gambarova NG, Gins MS (2008) Characteristics of oxidative stress of plants with C3 and C4 photosynthesis during salinization. Russ Agric Sci 34:77–80

Gapinska M, Sklodowska M, Gabara B (2008) Effect of short- and long-term salinity on the activities of antioxidative enzymes and lipid peroxidation in tomato roots. Acta Physiol Plant 30:11–18

Gill SS, Tuteja N (2010) Reactive oxygen species and antioxidant machinery in abiotic stress tolerance in crop plants. Plant Physioland Biochem 48(12):909–930

Golldack D, Li C, Mohan H, Probst N (2014) Tolerance to drought and salt stress in plant: unravelling the signalling networks. Frontiers Plant Sci 5

Grene R (2002) Oxidative stress and acclimation mechanisms in plants. In: Somerville CR, Meyerowitz EM (eds) The Arabidopsis Book. The American Society of Plant Biologists, Rockville, pp 1–20

Gupta R, Luan S (2003) Redox control of protein tyrosine phosphatases and mitogen activated protein kinases in plants. Plant Physiol 132:1149–1152

Halliwell B, Gutteridge JMC (1998) Free radicals in biology and medicine. Oxford University Press, Oxford

Hasanuzzaman M, Nahar K, Alam MM, Roychowdhury R, Fujita M (2013) Physiological, biochemical and molecular mechanisms of heat stress tolerance in plants. Int J Mol Sci 14:9643–9684

Hojati M, Modarres-Sanavy SAM, Karimi M, Ghanati F (2010) Responses of growth and antioxidant systems in *Carthamus tinctorius* L. under water deficit stress. Acta Physiol Plant 33:105–112

Holländer-Czytko H, Grabowski J, Sandorf I, Weckermann K, Weiler EW (2005) Tocopherol content and activities of tyrosine aminotransferase and cystine lyase in *Arabidopsis* under stress conditions. J Plant Physiol 162:767–770

Hong CY, Hsu YT, Tsai YC, Kao CH (2007) Expression of ASCORBATE PEROXIDASE 8 in roots of rice (*Oryza sativa* L.) seedlings in response to NaCl. J Exp Bot 58:3273–3283

Horemans N, Foyer CH, Asard H (2000) Transport and action of ascorbate at the plant plasma membrane. Trends Plant Sci 5:263–267

Hu X, Jiang M, Zhang J, Zhang A, Lin F, Tan M (2007) Calcium-calmodulin is required for abscisic acid-induced antioxidant defense and functions both upstream and downstream of H_2O_2 production in leaves of maize (*Zea mays*) plants. New Phytol 173:27–38

Igamberdiev AU, Seregelyes C, Manac N, Hill RD (2004) NADH- dependent metabolism of nitric oxide in alfalfa root cultures expressing barley hemoglobin. Planta 219:95–102

Ingram J, Bartels D (1996) The molecular basis of dehydration tolerance in plants. Annu Rev Plant Physiol Plant Mol Biol 47:377–403

Jain K, Kataria S, Guruprasad KN (2003) Changes in antioxidant defenses of cucumber cotyledons in response to UV-Band to the free radicals generating compound AAPH. Plant Sci 165:551–557

Jiang M, Zhang J (2003) Cross-talk between calcium and reactive oxygen species originated from NADPH oxidase in abscisic acid-induced antioxidant defence in leaves of maize seedlings. Plant Cell Environ 26:929–939

Joo JH, Wang S, Chen JG, Jones AM, Fedoroff NV (2005) Different signaling and cell death roles of heterotrimeric G protein a and b subunits in the *Arabidopsis* oxidative stress response to ozone. Plant Cell 17:957–970

Jurca ME, Bottka S, Fehér A (2008) Characterization of a family of Arabidopsis receptor-like cytoplasmic kinases (RLCK class VI). Plant Cell Rep 27(4):739–748

Kandziora-Ciupa M, Ciepal R, Nadgórska-Socha A, Barczyk G (2013) A comparative study of heavy metal accumulation and antioxidant responses in *VacciniummyrtillusL*. leaves in polluted and non-polluted areas. Environ Sci Pollut Res 20:4920–4932

Karuppanapandian T, Manoharan K (2008) Uptake and translocation of tri- and hexa-valent chromium and their effects on black gram (*Vigna mungo* L. Hepper cv. Co4) roots. J Plant Biol 51:192–201

Karuppanapandian T, Sinha PB, Kamarul Haniya A, Manoharan K (2006a) Differential antioxidative responses of ascorbate-glutathione cycle enzymes and metabolites to chromium stress in green gram (*Vigna radiata* L. Wilczek) leaves. J Plant Biol 49:440–447

Karuppanapandian T, Sinha PB, Premkumar G, Manoharan K (2006b) Chromium toxicity: Correlated with increased in degradation of photosynthetic pigments and total soluble protein and increased peroxidase activity in green gram (*Vigna radiata* L.) seedlings. J Swamy Bot-Cl 23:117–122

Kiffin R, Bandyopadhyay U, Cuervo AM (2006) Oxidative stress and autophagy. Antioxid Redox Signal 8:152–162

Koussevitzky S, Suzuki N, Huntington S, Armijo L, Sha W, Cortes D, Shulaev V, Mittler R (2008) Ascorbate peroxidase 1 plays a key role in the response of *Arabidopsis thaliana* to stress combination. J Biol Chem 283:34197–34203

Kovtun Y, Chiu W-L, Tena G, Sheen J (2000) Functional analysis of oxidative stress-activated mitogen-activated protein kinase cascade in plants. Proc Natl Acad Sci U S A 97:2940–2945

Kukreja S, Nandval AS, Kumar N, Sharma SK, Unvi V, Sharma PK (2005) Plant water status, H2O2 scavenging enzymes, ethylene evolution and membrane integrity of *Cicer arietinum* roots as affected by salinity. Biol Plant 49:305–308

Lin F, Ding H, Wang J, Zhang H, Zhang A, Zhang Y et al (2009) Positive feedback regulation of maize NADPHoxidase by mitogen-activated protein kinase cascade in abscisic acid signalling. J Exp Bot 60:3221–3238

Liu X, Hua X, Guo J, Qi D, Wang L, Liu Z (2008) Enhanced tolerance to drought stress in transgenic tobacco plants overexpressing *VTE1* for increased tocopherol production from *Arabidopsis thaliana*. Biotechnol Lett 30:1275–1280

Lopez-Huertas E, Charlton WL, Johnson B, Graham I, Baker A (2000) Stress induces peroxisome biogenesis genes. EMBO J 19(24):6770–6777

Lu ZQ, Liu D, Liu SK (2007) Two rice cytosolic ascorbate peroxidases differentially improve salt tolerance in transgenic *Arabidopsis*. Plant Cell Rep 26:1909–1917

Li J, Yang L, Jin D et al (2013) UV-B-induced photomorphogenesis in Arabidopsis. Protein Cell 4:485–492

Ma F, Lu R, Liu H, Shi B, Zhang J, Tan M et al (2012) Nitric oxide-activated calcium/calmodulin-dependent protein kinase regulates the abscisic acid-induced antioxidant defence in maize. J Exp Bot 63:4835–4847

Mafakheri A, Siosemardeh A, Bahramnejad B, Struik PC, Sohrabi Y (2010) Effect of drought stress on yield, proline and chlorophyll contents in three chickpea cultivars. Aus J Crop Sci 4:580–585

Marshall A, Aalen RB, Audenaert D, Beeckman T, Broadley MR, Butenko MA, Caño-Delgado AI, de Vries S, Dresselhaus T, Felix G (2012) Tackling drought stress: receptor-like kinases present new approaches. Plant Cell 24(6):2262–2278

Mhamdi A (2010) Catalase function in plants: a focus on Arabidopsis mutants as stress-mimic models. J Exp Bot 61:4107–4320

Miller G, Suzuki N, Rizhsky L, Hegie A, Koussevitzky S, Mittler R (2007) Double mutants deficient in cytosolic and thylakoid ascorbate peroxidase reveal a complex mode of interaction between reactive oxygen species, plant development, and response to abiotic stresses. Plant Physiol 144:1777–1785

Miller G, Suzuki N, Ciftci-Yilmaz S, Mittler R (2010) Reactive oxygen species homeostasis and signalling during drought and salinity stresses. Plant Cell Environ 33:453–467

Mishra A, Kumar S, Pandey AK (2013) Scientific validation of the medicinal efficacy of Tinospora cordifolia. Sci World J:292934

Mittler R (2002) Oxidative stress, antioxidants, and stress tolerance. Trends Plant Sci 7:405–410
Mittler R (2017) ROS are good. Trends Plant Sci 22(01):11–19
Morris ER, Walker JC (2003) Receptor-like protein kinases: the keys to response. Curr Opin Plant Biol 6(4):339–342
Mortensen A, Skibsted LH, Truscott TG (2001) The interaction of dietary carotenoids with radical species. Arch Biochem Biophys 385:13–19
Mullineaux PM, Rausch T (2005) Glutathione, photosynthesis and the redox regulation of stress-responsive gene expression. Photosyn Res 86:459–474
Munné-Bosch S, Alegre L (2000) Changes in carotenoids, tocopherols and diterpenes during drought and recovery, and the biological significance of chlorophyll loss in *Rosmarinus officinalis* plants. Planta 210:925–931
Nagamiya K, Motohashi T, Nakao K, Prodhan SH, Hattori E, Hirose S, Ozawa K, Ohkawa Y, Takabe T, Takabe T, Komamine A (2007) Enhancement of salt tolerance in transgenic rice expressing an *Escherichia coli* catalase gene, katE. Plant Biotechnol Rep 1:49–55
Nath K, Kumar S, Poudyal RS, Yang YN, Timilsina R, Park YS et al (2014) Developmental stage-dependent differential gene expression of superoxide dismutase isoenzymes and their localization and physical interaction network in rice (*Oryza sativa* L.). Genes Genom 36:45–55
Naya L, Ladrera R, Ramos J, González EM, Arrese-Igor C, Minchin FR, Becana M (2007) The response of carbon metabolism and antioxidant defenses of alfalfa nodules to drought stress and to the subsequent recovery of plants. Plant Physiol 144:1104–1114
Noctor G, Foyer CH (1998a) A re-evaluation of the ATP: NADPH budget during C3 photosynthesis. A contribution from nitrate assimilation and its associated respiratory activity? J Exp Bot 49:1895–1908
Noctor G, Foyer CH (1998b) Ascorbate and glutathione: keeping active oxygen under control. Ann Rev Plant Physiol Plant Mol Biol 49:249–279
Pan Y, Wu LJ, Yu ZL (2006) Effect of salt and drought stress on antioxidant enzymes activities and SOD isoenzymes of liquorice (*Glycyrrhiza uralensis* Fisch). J Plant Growth Regul 49:157–165
Pandey K, Mishra AK, Mishra A (2012) Antifungal and antioxidative potential of oil and extracts derived from leaves of Indian spice plant *Cinnamomum tamala*. Cell Mol Biol 58:142–147
Porcel R, Barea JM, Ruiz-Lozano JM (2003) Antioxidant activities in mycorrhizal soybean plants under drought stress and their possible relationship to the process of nodule senescence. New Phytol 157(1):135–143
Ramegowda V, Basu S, Krishnan A, Pereira A (2014) Rice growth under drought kinase is required for drought tolerance and grain yield under normal and drought stress conditions. Plant Physiol 166(3):1634–1645
Rentel MC, Lecourieux D, Ouaked F, Usher SL, Peterson L, Okamoto H, Knight H, Peck SC, Grierson CS, Hirt H, Knight MR (2004) OXI1 kinase is necessary for oxidative burst-mediated signaling in *Arabidopsis*. Nature 427:858–861
Rice-Evans CA, Burdon R (1993) Free radical-lipid interactions and their pathological consequences. Prog Lipid Res 32:71–110
Rizhsky L, Hallak-Herr E, Van Breusegem F, Rachmilevitch S, Barr JE, Rodermel S, Inze D, Mittler R (2002) Double antisense plants lacking ascorbate peroxidase and catalase are less sensitive to oxidative stress than single antisense plants lacking ascorbate peroxidase or catalase. Plant J 32:329–342
RoyChoudhury A, Basu S (2012) Ascorbate-Glutathione and plant tolerance to various abiotic stresses. In: Anjum NA, Umar S, Ahmad A (eds) Oxidative stress in plants: causes, consequences and tolerance. IK International Publishers, New Delhi, pp 177–258
RoyChoudhury A, Ghosh S (2013) Physiological and biochemical responses of mungbean (Vigna radiate L.Wilczek) to varying concentrations of cadmium chloride or sodium chloride. Unique J Pharm Biol Sci 1:11–21
RoyChoudhury A, Basu S, Sarkar SN, Sengupta DN (2008) Comparative physiological and molecular responses of a common aromatic indica rice cultivar to high salinity with non-aromatic indica rice cultivars. Plant Cell Rep 27:1395–1410

RoyChoudhury A, Pradhan S, Chaudhuri B, Das K (2012a) Phytoremediation of toxic metals and the involvement of *Brassica* species. In: Anjum NA, Pereira ME, Ahmad I, Duarte AC, Umar S, Khan NA (eds) Phytotechnologies: remediation of Environmental Contaminants. CRC press, Taylor and Francis Group, Boca Raton, pp 219–251

RoyChoudhury A, Das K, Ghosh S, Mukherjee RN, Banerjee R (2012b) Transgenic Plants: benefits and controversies. J Bot Soc Bengal 66:29–35

RoyChoudhury A, Basu S, Sengupta DN (2012c) Antioxidants and stress- related metabolites in the seedlings of two indica rice varieties exposed to cadmium chloride toxicity. Acta Physiol Plant 34:835–847

Sairam RK, Deshmukh PS, Saxena DC (1998) Role of antioxidant systems in wheat genotypes tolerance to water stress. Biol Plant 41:387–394

Selote DS, Khanna-Chopra R (2010) Antioxidant response of wheat roots to drought acclimation. Protoplasma 245:153–163

Setterdahl AT, Chivers PT, Hirasawa M, Lemaire SD, Keryer E, Miginiac-Maslow M, Kim SK, Mason J, Jacquot JP, Longbine CC, de Lamotte-Guery F, Knaff DB (2003) Effect of pH on the oxidation-reduction properties of thioredoxins. Biochemistry 42:14877–14884

Shalata A, Neumann PM (2001) Exogenous ascorbic acid (vitamin C) increases resistance to salt stress and reduces lipid peroxidation. J Exp Bot 52:2207–2211

Shao HB, Liang ZS, Shao MA, Sun Q (2005) Dynamic changes of anti- oxidative enzymes of 10 wheat genotypes at soil water deficits. Colloids Surf B Biointerfaces 42:187–195

Shao HB, Chu LY, Shao MA, Jaleel CA, Mi HM (2008) Higher plant antioxidants and redox signaling under environmental stresses. C R Biol 331(6):433–441

Sharma P, Dubey RS (2004) Ascorbate peroxidase from rice seedlings: properties of enzyme isoforms, effects of stresses and protective roles of osmolytes. Plant Sci 167:541–550

Sharma P, Dubey RS (2005a) Drought induces oxidative stress and enhances the activities of antioxidant enzymes in growing rice seedlings. Plant Growth Regul 46:209–221

Sharma P, Dubey RS (2005b) Modulation of nitrate reductase activity in rice seedlings under aluminium toxicity and water stress: role of osmolytes as enzyme protectant. J Plant Physiol 162:854–864

Shi B, Ni L, Zhang A, Cao J, Zhang H, Qin T et al (2012) OsDMI3 is a novel component of abscisic acid signaling in the induction of antioxidant defense in leaves of rice. Mol Plant 5:1359–1374

Shinozaki K, Yamaguchi-Shinozaki K (2007) Gene networks involved in drought stress response and tolerance. J Exp Bot 58(2):221–227

Shinozaki K, Yamaguchi-Shinozaki K, Seki M (2003) Regulatory network of gene expression in the drought and cold stress responses. Curr Opin Plant Biol 6:410–417

Shiu SH, Bleecker AB (2001) Receptor-like kinases from Arabidopsis form a monophyletic gene family related to animal receptor kinases. Proc Nat Acad Sci U S A 98(19):10763–10768

Shiu SH, Bleecker AB (2003) Expansion of the receptor-like kinase/Pelle gene family and receptor-like proteins in Arabidopsis. Plant Physiol 132:530–543

Simova-Stoilova L, Vaseva I, Grigorova B, Demirevska K, Feller U (2010) Proteolytic activity and cysteine protease expression in wheat leaves under severe soil drought and recovery. Plant Physiol Biochem 48:200–206

Smirnoff N (2000) Ascorbic acid: metabolism and functions of a multi-facetted molecule. Curr Opin Plant Biol 3:229–235

Sreedevi S, Krishnan PN (2012) Detoxification of heavy metal induced oxidative stress in plants, enzymatic and non-enzymatic machanisms-amini review. In: Anjum NA, Umar S, Ahmad A (eds) Oxidative stress in plants: causes, consequences and tolerance. IK International Publishing House, New Delhi, pp 291–307

Srivastava AK, Bhargava P, Rai LC (2005) Salinity and copper-induced oxidative damage and changes in antioxidative defense system of Anabaena doliolum. W J Microb Biotechnol 22:1291–1298

Stoop JMH, Williamson JD, Pharr DM (1996) Mannitol metabolism in plants: a method for coping with stress. Trends Plant Sci 1:139–155

Sun YY, Sun YJ, Wang MT, Li XY, Guo X, Hu R, Ma J (2010) Effects of seed priming on germination and seedling growth of rice under water stress. Acta Agron Sin 36(11):1931–1940

Tarczynski MC, Jensen RG, Bohnert HJ (1993) Stress protection in transgenic tobacco by production of the osmolyte mannitol. Science 259:508–510

Teixeira FK, Menezes-Benavente L, Galvão VC, Margis R, Margis-Pinheiro M (2006) Rice ascorbate peroxidase gene family encodes functionally diverse isoforms localized in different subcellular compartments. Planta 224:300–314

Tippmann HF, Schluter U, Collinge DB (2006) Common Themes in Biotic and Abiotic Stress Signalling in Plants. Floric Ornam Plant Biotechnol 3:52–67

Torres MA (2010) ROS in biotic interactions. Physiol Plant 138:414–429

Tsuchiya H, Linuma M (2000) Reduction of membrane fluidity by antibacterial sophoraflavanone G isolated from Sophoraexigua. Phytomedicine 7(2):161–165

Turan O, Ekmekci Y (2011) Activities of photosystem II and antioxidant enzymes in chickpea (*Cicer arietinum* L.) cultivars exposed to chilling temperatures. Acta Physiol Plant 33:67–78

Ushimaru T, Nakagawa T, Fujioka Y, Daicho K, Naito M, Yamauchi Y, Nonaka H, Amako K, Yamawaki K, Murata N (2006) Transgenic Arabidopsis plants expressing the rice dehydroascorbate reductase gene are resistant to salt stress. J Plant Physiol 163:1179–1184

Vaidyanathan H, Sivakumar P, Chakrabarty R, Thomas G (2003) Scavenging of reactive oxygen species in NaCl-stressed rice (*Oryza sativa* L.)—differential response in salt-tolerant and sensitive varieties. Plant Sci 165:1411–1418

Vellosillo T, Vicente J, Kulasekaran S, Hamberg M, Castresana C (2010) Emerging complexity in reactive oxygen species production and signaling during the response of plants to pathogens. Plant Physiol 154:444–448

Verbruggen N, Hermans C (2008) Proline accumulation in plants: a review. Amino Acids 35:753–759

Vij S, Giri J, Dansana PK, Kapoor S, Tyagi AK (2008) The receptor-like cytoplasmic kinase (OsRLCK) gene family in rice: organization, phylogenetic relationship, and expression during development and stress. Mol Plant 1(5):732–750

Wang Y, Ying Y, Chen J, Wang XC (2004) Transgenic Arabidopsis overexpressing Mn-SOD enhanced salt-tolerance. Plant Sci 167:671–677

Wang Y, Wisniewski M, Meilan R, Uratsu SL, Cui MG, Dandekar A, Fuchigami L (2007) Ectopic expression of Mn-SOD in *Lycopersicon esculentum* leads to enhanced tolerance to salt and oxidative stress. J Appl Horticul 9:3–8

Wang L, Zhou Q, Ding L, Sun Y (2008) Effect of cadmium toxicity on nitrogen metabolism in leaves of *Solanum nigrum* L. as a newly found cadmium hyperaccumulator. J Hazard Met 154:818–825

Wang R, Liu S, Zhou F, Ding C, Hua C (2014) Exogenous ascorbic acid and glutathione alleviate oxidative stress induced by salt stress in the chloroplasts of *Oryza sativa* L. Z Naturforsch C 69:226–236

Wen F, Qin T, Wang Y, Dong W, Zhang A, Tan M et al (2015) OsHK3 is a crucial regulator of abscisic acid signaling involved in antioxidant defense in rice. J Integr Plant Biol 57:213–228

Witzelei K, Weidner A, Surabhi GK (2009) Bo¨ rner A, Mock HP. Salt stress-induced alterations in the root proteome of barley genotypes with contrasting response towards salinity. J Exp Bot 60(12):3545–3557

Wu L, Zhang Z, Zhang H, Wang XC, Huang R (2008) Transcriptional modulation of ethylene response factor protein JERF3 in the oxidative stress response enhances tolerance of tobacco seedlings to salt, drought, and freezing. Plant Physiol 148:1953–1963

Xing Y, Jia W, Zhang J (2008) AtMKK1 mediates ABA-induced CAT1 expression and H2O2 production via AtMPK6-coupled signaling in Arabidopsis. Plant J 54:440–451

Xiong L, Schumaker KS, Zhu JK (2002) Cell signaling during cold, drought, and salt stress. Plant Cell 14:165–183

Xu C, Sullivan JH, Garrett WM, Caperna TJ, Natarajan S (2008) Impact of solar Ultraviolet-Bonproteome in soybean lines differing in flavonoid contents. Phytochemistry 69:38–48

Yamaguchi-Shinozaki K, Shinozaki K (2005) Organization of cis-acting regulatory elements in osmotic- and cold-stress-responsive promoters. Trends Plant Sci 10:88–94

Yamaguchi-Shinozaki K, Shinozaki K (2006) Transcriptional regulatory networks in cellular responses and tolerance to dehydration and cold stresses. Ann Rev in Plant Biol 57:781–803

Yan J, Guan L, Sun Y, Zhu Y, Liu L, Lu R et al (2015) Calcium and ZmCCaMK are involved in brassinosteroid-induced antioxidant defense in maize leaves. Plant Cell Physiol 56:883–896

Yang Z, Wu Y, Li Y, Ling HQ, Chu C (2009) OsMT1a, a type 1 metallothionein, plays the pivotal role in zinc homeostasis and drought tolerance in rice. Plant Mol Biol 70:219–229

Yang CJ, Zhang C, Lu YN, Jin JQ, Wang XL (2011) The mechanisms of brassinosteroids' action: from signal transduction to plant development. Mol Plant 4:588–600

Ye N, Zhu G, Liu Y, Li Y, Zhang J (2011) ABA controls H_2O_2 accumulation through the induction of OsCATB in rice leaves under water stress. Plant Cell Physiol 52:689–698

Zhang A, Jiang M, Zhang J, Tan M, Hu X (2006) Mitogen-activated protein kinase is involved in abscisic acid-induced antioxidant defense and acts downstream of reactive oxygen species production in leaves of maize plants. Plant Physiol 141:475–487

Zhang A, Jiang M, Zhang J, Ding H, Xu S, Hu X et al (2007) Nitric oxide induced by hydrogen peroxide mediates abscisic acid-induced activation of the mitogen-activated protein kinase cascade involved in antioxidant defense in maize leaves. New Phytol 175:36–50

Zhang H, Ni L, Liu Y, Wang Y, Zhang A, Tan M et al (2012a) The C_2H_2-type zinc finger protein ZFP182 is involved in abscisic acid-induced antioxidant defense in rice. J Integr Plant Biol 54:500–510

Zhang L, Li Y, Lu W, Meng F, Wu CA, Guo X (2012b) Cotton GhMKK5affects disease resistance, induces HR-like cell death, and reduces the tolerance to salt and drought stress in transgenic *Nicotiana benthamiana*. J Exp Bot 63:3935–3951

Zhang Z, Zhang Q, Wu J, Zheng X, Zheng S, Sun X et al (2013) Gene knockout study reveals that cytosolic ascorbate peroxidase2 (OsAPX2) plays a critical role in growth and reproduction in rice under drought, salt and cold stresses. PLoSONE 8:e57472. https://doi.org/10.1371/journal.pone.0057472

Zhang H, Liu Y, Wen F, Yao D, Wang L, Guo J et al (2014) A novel rice C2H2-type zinc finger protein, ZFP36, is a key player involved in abscisic acid- induced antioxidant defence and oxidative stress tolerance in rice. J Exp Bot 65:5795–5809

Zhu J-K (2000) Genetic analysis of plant salt tolerance using *Arabidopsis thaliana*. Plant Physiol 124:941–948

Zhu J-K (2002) Salt and drought stress signal transduction in plants. Annu Rev Plant Biol 53:247–273

Zhu JY, Sae-Seaw J, Wang ZY (2013a) Brassinosteroid signalling. Development 140:1615–1620

Zhu Y, Zuo M, Liang Y, Jiang M, Zhang J, Scheller HV et al (2013b) MAP65-1 a positively regulates H_2O_2 amplification and enhances brassinosteroid-induced antioxidant defence in maize. J Exp Bot 64:3787–3802

Zlatev ZS, Lidon FC, Ramalho JC, Yordanov IT (2006) Comparison of resistance to drought of three bean cultivars. Biol Plant 50:389–394

ROS in Aging and Senescence

3

Abstract

As an unavoidable consequence of aging and natural course of senescence, disruption of redox homeostasis due to over-accumulation of ROS (reactive oxygen species) in plant cell is observed. Plants have evolved an array of self-protective defensive tools to oppose loss of redox homeostasis due to stress-induced aging and also natural course of senescence. However, it is becoming evident that ROS, which are generated during aging and natural course of senescence, are recognized by plant as a signaling agent for triggering responses. In fact, one of the earliest events upon recognition of an unfavorable environmental cue and infection is the accumulation of reactive oxygen species (ROS). The tissue necrosis triggered by reactive oxygen species (ROS) during biotic stress increases host susceptibility to necrotrophic but resistance to biotrophic pathogen. Strong evidences corroborate the view that ROS serve as a signaling agent in a systemic signaling network in plant cell leading to the expression of defense genes and triggering hypersensitive response (HR). Avirulent pathogens often induce an oxidative burst exhibiting a biphasic ROS accumulation. A range of enzyme systems like RBOH (NADPH oxidase) and SOD (superoxide dismutase) have been implicated in ROS interaction following pathogen recognition and PCD. Several PGRs like salicylic acid, jasmonic acid, and ethylene can influence initiation of ROS and antioxidants, thereby influencing resistance or susceptibility of plants to pathogens and cell death. In this chapter an effort has been made to describe the implications of ROS in cellular senescence, highlighting their signaling role. Special emphasis is given on the role of ROS in programmed cell death associated with hypersensitive response (HR) in plants.

Keywords

Oxidative stress · Programmed cell death · ROS signaling · Hypersensitive reaction · Redox homeostasis

S. Bhattacharjee, *Reactive Oxygen Species in Plant Biology*,
https://doi.org/10.1007/978-81-322-3941-3_3

3.1 Introduction

Even though aerobic metabolism is energetically competent, the occurrence of molecular oxygen in the cellular environment poses an invariable oxidative hazard or threat to all important classes of cellular macromolecules, thereby influencing vital life processes. An inevitable result of chloroplast, mitochondria, and plasma membrane-linked ETS (electron transport system) is the spilling of electrons onto molecular oxygen in plant cells, with the production of partially reduced toxic ROS (Asada 1994; Fridovich 1995; Varnova et al. 2002; Arora et al. 2002; Bhattacharjee 2005). Impositions of any kind of stress, both abiotic and biotic, further disrupt the redox status of the cell, causing accumulation of prooxidants, and accelerate oxidative damage (Arora et al. 2002). Redox status of the cell also determine natural event of senescence, as evidenced by the elevated titer of ROS, and associated oxidative damages to thylakoid and inner mitochondrial membranes (Thompson et al. 1987; Vacca et al. 2004).

In fact, any state in which cellular redox homeostasis of the cell is disturbed due to an imbalance in which the redox steady state of the cell is altered significantly in the direction of prooxidants can be defined as "oxidative stress." The ROS normally produced by the cell and capable of inducing oxidative stress when accumulated include superoxide ($O_2^{\cdot -}$), perhydroxy radical ($HO_2^{\cdot -}$), hydrogen peroxide (H_2O_2), hydroxyl radical ($OH^{\cdot}$), alkoxy radical ($RO^{\cdot}$), peroxy radical ($ROO^{\cdot}$), singlet oxygen ($^1O_2$), organic hydroperoxide (ROOH), etc. Therefore, because of the highly cytotoxic and reactive nature of ROS, their turnover and accumulation must be under tight control. Plants possess very efficient antioxidant defense systems that not only allow scavenging of ROS and protection of plants from oxidative damage but also buffer the endogenous titer. The distinct subcellular localization and biochemical properties of antioxidant enzymes, their differential inducibility at the enzyme and gene expression level, and the plethora of nonenzymatic scavengers render the antioxidant systems a very versatile and flexible unit that can control ROS accumulation.

Many environmental stresses including infection result in increased generation of ROS. This oxidative burst often leads to the stimulation of defensive mechanisms, including changes in gene expression, which lead the recovery of redox equilibrium and revival from damage or toxicity. However, little is known about the signaling procedures that is associated with the perception of oxidative burst to gene expression leading to stress amelioration and defense. Recent works also suggest that oxidative burst (originating during pathogenesis and abiotic stress) can act as signaling event, which in turn can trigger cell death in plants. The involvement of superoxide in plant hypersensitive responses and hydrogen peroxide in plant cell death mediated by pathogens are reported. However we do not have a clear idea about the exact role of these two ROS in the process of cell death in plants.

3.2 Implication of ROS in Plant Senescence

Plant senescence is a complex deteriorative process set into motion according to the hereditary instruction leading to the termination of functional life of whole organism, an organ, tissue, or even a cell. It is regulated by endogenous autonomous factors (including age, reproductive development, and levels of growth factors) and by environmental signals (including photoperiod and other environmental odds or stresses such as drought, heat, chilling, nutrient deficiency, wounding, heavy metals, shading, etc.) (Gan and Amasino 1997). The production of ROS is one of the initial responses of plant cells under abiotic stresses and senescence (Lee et al. 2012). In plants, ROS are formed as by-products of aerobic energy metabolism and also as a result of exposure to various stresses, both biotic and abiotic (Selote and Khanna-Chopra 2006). Under normal homeostatic conditions, the ROS turnover is balanced in cells and is maintained at low levels by antioxidant defense system. This redox homeostasis can be disturbed either by the reduction of antioxidants or the excess production of ROS, leading to oxidative stress, and consequently cause oxidative damage to cellular macromolecules and membranes (Lushchak 2011). Plants have evolved mechanism that afford them protection against oxidative deterioration, which involves less ROS production coupled with upregulation of an efficient antioxidant defense system (2007) as a part of activation of different redox signaling pathways (Khanna-Chopra and Selote 2007; Wang et al. 2014).

Increased level of ROS in senescing tissue or cell could occur through either enhanced production of ROS or decline in the efficiency of antioxidative defense mechanisms that normally afford protection against oxidative injury (Thompson et al. 1998; Bhattacharjee 2005). In fact senescence encompasses ordered degenerative processes which unlike the process of aging are instigated by hereditary instruction of the cell. Aging, on the contrary, is wear and tear that accumulates due to imposition of unfavorable environmental cues and enhances the probability of senescence. So, the loss of redox homeostasis is one of the prime markers of the senescing cell, and the manifestation of this state of the cell ranges from membrane damage, metabolic impairment to cascades of oxidative deterioration including genomic lesion (Fig. 3.1). The important role of ROS in signaling during senescence has been demonstrated in many studies. It has been shown that ROS modulate the activity of key signaling compounds and have an important role during early responses to senescence (Jajic et al. 2015). Despite the importance of ROS in several cellular processes, our knowledge of the mechanism of their action during senescence is still limited. In this chapter we will try to summarize the latest progress on the roles of reactive oxygen species during senescence.

Paradoxically, senescence is the terminal developmental process and hence is an integral part of life. Apart from termination of functional life, cellular senescence is essential for regulating and coordinating growth and development of plants, by sustaining tissue and organ homeostasis in association with cell proliferation and differentiation. The wide diversity of cellular senescence types, reported in the literature, was mostly categorized into two categories: necrosis and apoptosis. Differences between these two types were based on the presence or absence of

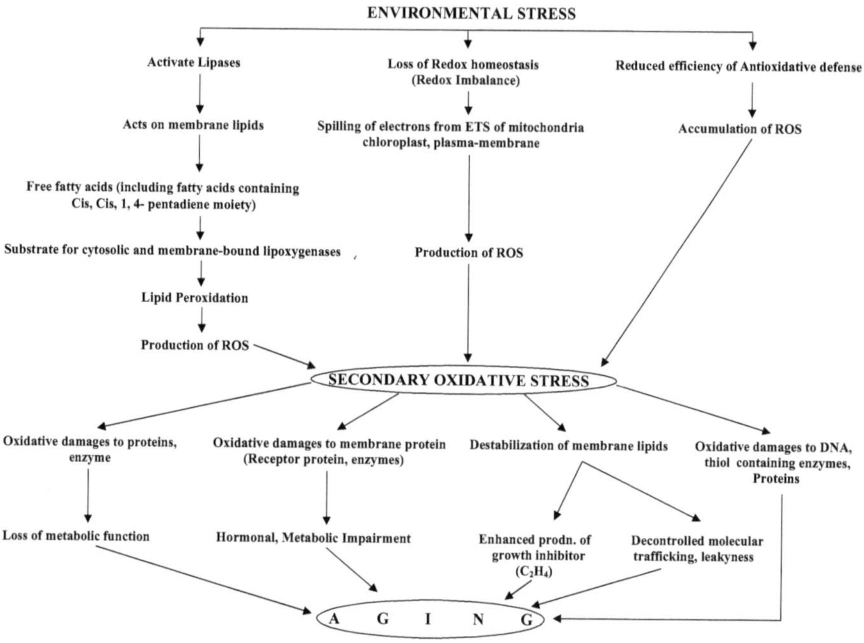

Fig. 3.1 Loss of redox homeostasis and the manifestation of oxidative deterioration during cellular senescence

specific molecular, biochemical, and metabolic hallmarks, such as DNA laddering, cytochrome c release, caspase involvement, ATP depletion, redox imbalance, metabolic dysfunction, loss of ion homeostasis, cytoplasmic swelling, and loss of membrane integrity and deregulation of membrane trafficking (Pennell and Lamb 1997). Nonetheless, over the last decade, this arbitrary separation had clearly become too simplistic, and a more precise account of plant cell death was established (van Doorn and Woltering 2005). To avoid ambiguity of terminology, it may be proposed that necrosis mainly represents cell death caused by extrinsic factors, such as phytotoxic accumulation of specific molecules after a traumatic stress event or a resultant of aging induced by unfavorable external conditions. Thus, plant necrotic cell death is haphazard and unregulated and a consequence of irreversible damages. Progressive loss of membrane integrity with decontrolled membrane trafficking results in swelling of the cytoplasm and cellular lysis. The term programmed cell death (PCD), on the other hand, describes any form of cell death involving a series of molecular, biochemical, and cellular orderly events mediated by intracellular death programs that are set into motion according to genomic instruction, regardless of the trigger or the characteristics it exhibits (Jacobson et al. 1997; Dangl et al. 2000). Although the necrosis and PCD are quite well defined in animals, in plants there seems to have common characteristics between the biochemical and molecular traits of necrosis and PCD, which makes them difficult to differentiate. In plants, PCD can occur mainly during programmed developmental processes and as a response to biotic stress conditions. For example, PCD under stress has been illustrated during hypoxia (Drew et al. 2000), high and low temperature (Vacca et al. 2004), ozone exposure, etc. (Langebartels et al. 2002). PCD associated with the development of plants has been reported during the formation and differentiation of vascular bundles, particularly the tracheary elements (Kuriyama and Fukuda 2002), development and germination of seeds (Souter and Lindsey 2000; Young and Gallie 2000), as well as different forms of senescence processes (Gunawardena et al. 2004). However, the plant senescence events remain best described during incompatible plant-pathogen interactions that form the foundation for the hypersensitive response (Pennell and Lamb 1997). This multitude of plant PCD events undoubtedly illustrates a practical analogy between cell deaths across plant kingdoms. These include sculpting and deleting structures during development, eliminating cells to control cell quality and quantity after suffering damages, and producing differentiated cells without organelles (Jacobson et al. 1997). But, the question arises whether there exist more than just a functional similarity between plant and animal PCD? In animals the role of ROS as mediator of the process has been elucidated long back (Jabs 1999). In contrast, the role of ROS or redox regulation during plant PCD has for a long time remained rather hypothetical. Both the lack of sensitive and reliable analytical detection methods and the inherent toxic nature of ROS masked their underlying redox signaling function during senescence. However, in last decade, the prominent role of ROS has been exposed in the induction, signaling, and implementation of plant cell death. In this regard, the initial physiological and molecular evidence for ROS signaling and redox regulation of plant PCD has been established genetically with the detection of *Arabidopsis* mutants incompetent either of arresting oxidative stress-driven PCD or of ensuing PCD in spite of high oxidant levels. In contemporary times, genes associated with ROS perception and redox signaling associated with senescence have been identified, and the characterization of other factors are in the way, revealing the complicated redox signaling network of plant cell death.

3.3 Implication of ROS and Oxidative Stress in Programmed Cell Death of Plant

Selective elimination of unwanted cells in multicellular organism is the basic objective of PCD. The process can be characterized by consistent and inflexible involvement of cells own machinery as an "executioner," to instigate and mediate the suicidal process. Oxidative stress as signaling event is shown to trigger cell death in plants. The involvement of superoxide in plant HR and hydrogen peroxide in cell death mediated by pathogen is reported (Krishnamurthy et al. 1999). Though the precise role of ROS in the entire process of plant cell death is not very apparent, but in many cases, H_2O_2 is likely to persuade DNA lesion and influx of extracellular calcium, associated with PCD. ROS H_2O_2 has been reported not only in cells subjected to biotic and abiotic stress but is also steadily produced when there is an activity of peroxidase and polyamine oxidase, associated with the lignification and cross-linking of cell wall (Krishnamurthy et al. 1999).

In fact, a number of different categories of cells undergo programmed cell death, viz.:

(i) Cells that is damaged and unable to function anymore.
(ii) Cells that are subject to hypersensitive response.
(iii) Cells that have served their function.
(iv) Cells that are not needed from the beginning.
(v) Cells committed to die during differentiation.
(vi) Cells that are produced in excess.
(vii) Cells present in wrong sites.
(viii) Cells which through their death give rise to disease.

Of these categories, the role of ROS as an inducer of PCD is evident for cells subjected to hypersensitive response (HR) due to pathogenic and other abiotic stresses such as temperature, salt, water, UV, heavy metals, etc. Other category of cells where the role of ROS is apparent in mediating the cell death is the damaged cells that are unable to function.

The molecular mechanism of PCD in plants has been studied mostly related to pathogenesis and pathogen-induced oxidative burst. As expected, the accumulating evidence points to a complex signal transduction pathway, resembling the mammalian apoptotic pathway. For example, cysteine proteases play a major role in early steps of PCD in plants and animals, although the specific recognition sequence of protease cleavage may have diverged. Also the activation of protein kinases is central to both animals and plant system. The signaling pathway leading to the formation of apoptotic lesions in the *Arabidopsis* mutant (lsd1) has been partially elucidated. The biochemical analysis of the leaves of mutant plant implicated the superoxide metabolism as the key process in the initiation of PCD in this plant (Jabs et al. 1996).

One of the downstream cellular responses of different organisms to oxidative stress, particularly due to the generation of H_2O_2, is triggering programmed cell death. The death of the host cells during infection plays a major role in limiting the infection (hypersensitive response), as discussed earlier. Hypersensitive response (HR) is thought to be instrumental in limiting the spreading of pathogens from initial infection sites to neighboring cells, creating a gradient of H_2O_2 (Levine et al. 1994). Cell suicide also eliminates ROS-damaged cells, which would create a weakened area, susceptible to later attack by pathogens. Because of the highly permeable and moderately stable nature with low reactivity, H_2O_2 can diffuse away from the infected area to neighboring cells, creating a gradient of H_2O_2. At some point the concentration of H_2O_2 in neighboring cells reaches the level below the threshold concentration of H_2O_2 required for PCD. Moreover, to prevent the formation of a circle of weakened cells, the plant responds to the lower concentration of H_2O_2 by upregulating antioxidant response.

3.3.1 Tilting ROS Homeostasis to Induce Signaling During Senescence

Reactive oxygen species are natural by-products of aerobic metabolism involving redox reactions. Indeed, ROS are produced as a consequence of redox reactions associated with the most of the essential energy-generating processes. The redox cascades of chloroplasts, peroxisomes, and mitochondria are the main organelles for the generation of ROS in plant cells (Vranová et al. 2002; Apel and Hirt 2004; Bhattacharjee 2005). In spite of strong integration of the ETCs with efficient antioxidant defense system in these organelles, subtle changes in ROS homeostasis are inevitable due to fluctuating environmental conditions and developmental cues (Foyer and Noctor 2005). When the changes in endogenous concentrations of ROS are relatively minute, the competence of antioxidant defense is adequate to reset the original equilibrium between ROS production and scavenging, thus restoring redox homeostasis. Like unfavorable environmental conditions and infections, senescing condition also triggers loss of redox equilibrium, potentially leading to oxidative stress. In fact, a three- to tenfold augmentation in ROS levels has been calculated both under severe stress and senescing conditions (Polle 2001; Bhattacharjee 2005; Bhattacharjee et al. 2012). Thus momentary oxidative burst and a following temporary shift in the intracellular redox state are common features of both stress responses and senescence (Mittler et al. 2004). Plant approaching senescence significantly reduces the rate of PCRC, potentially steering the photosystems toward overproduction of ROS and inducing oxidative stress (Foyer and Noctor 2005). Under such conditions, various ROS react indiscriminately, with greater kinetics with almost all cellular components, and provoke destructive oxidative protein modifications, lipid peroxidations, mutagenic DNA strand breaks, purine oxidations, and unwanted protein-DNA cross-links (Beckman and Ames 1997). Peroxidation of membrane lipid takes place, both enzymatically and nonenzymatically, when above-threshold ROS levels are reached. These incidences during senescence is not only causing dismantling of organelle structure but directly impacting usual cellular metabolic homeostasis through the production of lipid-derived radicals (Montillet et al. 2005). As a result, the cellular oxidative damage resulting from tilting the balance of ROS levels shows some of the characteristic traits of necrosis. It is primarily because phytotoxic levels of prooxidants erratically attack cellular constituents, leading to loss of membrane architecture and cell lyses.

The first sign for participation of ROS or prooxidants in PCD was based chiefly on spatiotemporal association between significantly enhanced ROS levels and cellular senescence. That ROS can induce signaling associated with plant PCD can be demonstrated from the observation that H_2O_2-induced plant cell death could be blocked by the application of cycloheximide and specific protease inhibitors (Levine et al. 1994). Since then, there has been a growing evidence for a key role for ROS as elicitor of PCD. In this regard, though the initial results were obtained by exogenous use of ROS H_2O_2, but of late, transgenics and mutants with reduced levels of cellular antioxidants and lack of ability to hinder propagation of ROS-driven cell death elegantly demonstrated the important role of ROS in plant PCD events

(Lorrain et al. 2003). The best studied example of ROS-derived PCD is those following the characteristic biphasic oxidative burst during hypersensitive reaction (HR). The resulting cell death is demonstrated by the appearance of distinct cellular lesions that are generally preceded by the manifestation of several features of PCD, such as DNA laddering, condensation of chromatin, release of cytochrome c, etc. (Lam 2004). Furthermore, these characteristics can be partially reverted by administrating either elevated concentrations of antioxidants or inhibitors of protein synthesis associated with known signal transducing components, such as kinases, phosphatases, or even growth regulators. In view of that, genetically modified plants with low or high levels of enzymatic antioxidant competence, such as SOD, CAT, APOX, GR, etc., can also display an altered response to stress-driven PCD, again demonstrating the importance of a firmly orchestrated redox homeostasis of the cell in the process senescence (Mittler et al. 1999).

3.4 ROS Signaling Associated with Programmed Cell Death of Plants

Out of several ROS, H_2O_2 has been implicated as the most important oxidant mediating PCD, which occurs mainly during HR and also under suspension cultures of plants (Pennell and Lamb 1997). Application of harpin in *Arabidopsis* suspension cultures induced PCD, and this result is mediated at least in part by ROS H_2O_2 (Desikan et al. 2001). Similarly, the exogenous application of H_2O_2 can also mediate cell death in a dose- and time-dependent manner. The requirement of de novo protein synthesis, both transcription and translation, under the influence of ROS during its "presentation time" further proves its controlled orchestrated action during the process senescence.

Evidences associated with H_2O_2-mediated cell signaling processes leading to senescence are scanty. However, some evidences in support of ROS signaling associated with PCD do exist. H_2O_2-induced PCD in soybean cultures found to exploit Ca^{2+} signaling (Apel and Hirt 2004). The ROS-dependent Ca^{2+} influx and protein phosphorylation are found to be closely connected during PCD of soybean-cultured cells. ROS signaling results in increased cytosolic calcium level, which in turn activated a MAP kinase, just like abiotic and biotic stress responses. Harpin-induced rapid upregulation in activity of a MAP kinase like protein kinase with 44 KDa molecular mass was found to have similar mechanism of action with H_2O_2-induced activation of MAP kinases. Nevertheless, it remains still illusive, whether or not the upregulation of protein kinase activity is entirely indispensable for H_2O_2-induced gene expression associated with PCD (Fig. 3.2).

Necrosis is another form of cell death that does occur in plants apart from PCD. However, in contrast to PCD, which is genetically synchronized and controlled, necrosis is found to be a consequence of harsh and persistent disturbances and considered not to be genetically orchestrated (Pennel and Lamb 1997). Both the processes, PCD and necrosis, have been found to be just two discrete ends of the identical process of cell death that can be initiated and mediated by the same ROS

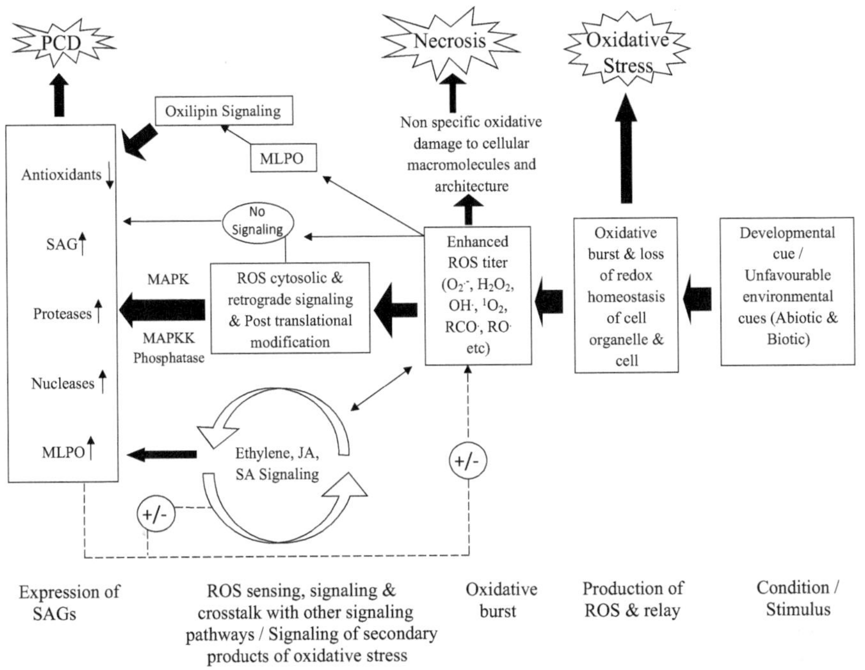

Fig. 3.2 An integrative model showing ROS-dependent cellular senescence of plant. Developmental cues and environmental stress (both abiotic and biotic) causes loss of redox homeostasis by upregulating accumulation of ROS, which might further got enhanced by various relay and production centers. Sustained ROS production and inefficiency of antioxidative defense caused the accumulation of phytotoxic levels of ROS leading to necrotic cell death. Transient climacteric rise in the level of ROS may initiate signal transduction cascades involving cross talk with other phytohormones. This cross talk can modulate the ensuing response through more specific transducer sensors and targets toward ROS-dependent expression of senescence-associated genes. MAPK-driven phosphorylation cascades and other regulatory posttranslational modifications, such as protein oxidation and nitrosylation, might be involved in ROS-dependent cell death pathways. *ETH* ethylene, *exe1* executer1, *JA* jasmonic acid

signal or decontrolled oxidative stress (Jabs 1999). H_2O_2 is biochemically capable to set in motion a cellular senescence by instigating as signaling molecule through a specific signal transduction event rather than killing cells by causing oxidative deteriorations that came from soybean cell culture experiments, in which short pulse of ROS is found to be capable of catalyzing the event (Levine et al. 1994). Therefore, low-dose exogenous application of H_2O_2 (< 5 mM) is capable of initiating an active cellular senescence pathway, which needs both DNA and protein syntheses in suspension cultures of *Arabidopsis* (Levine et al. 1994). Several works indicated that H_2O_2 initiates a cellular senescence program via cross talking, both feed forward and backward, with other growth regulators or signaling compounds, such as ABA, ethylene, jasmonic acid, and salicylic acid (Overmyer et al. 2005). Jabs et al. (1996) demonstrated that out of several ROS, $O_2^{\cdot-}$ commence a cellular senescence mechanism in *Arabidopsis* lsd1 mutant, vouching strong molecular genetics evidences for the role of this ROS in plant cell death. This *Arabidopsis* mutant (*lsd1*) grown under long days spontaneously forms necrotic lesion on leaves and consequently became completely lethal (Overmyer et al. 2005). Moreover, the spreading zone of cell death found to accumulate the persistantly significant amount of $O_2^{\cdot-}$. So, at least in this type of cellular senescence, the ROS $O_2^{\cdot-}$ appear to be significant death signal mediating the episode and can be monitored via a "rheostat" LCD1. However, when compared between $O_2^{\cdot-}$ and H_2O_2 in cell death activation mechanism, both the ROS have been found to be unquestionably involved in the signaling process and mediate genetically regulated senescence processes in the plant.

PCD occurring during hypoxia in plants is also characterized by phosphorylation/dephosphorylation events (Pennell and Lamb 1997). During the hypersensitive reactions in aleurone cells of seed, the supply of okadaic acid (protein phosphatase inhibitor) has been found to hinder cell death, vouching once again the involvement of cell signaling during the process (Kuo et al. 1996). Therefore, the HR reaction associated with pathogenesis of plant cells exhibits cell signaling with protein phosphorylation/dephosphorylation in the downstream (Mittler et al. 1999). This event of phosphorylation/dephosphorylation of target proteins is also associated with ROS signaling network (Mittler et al. 1999). So, whether the pattern of ROS signaling, involving phosphorylation and dephosphorylation, is similar to that of PCD or independent one, needs to be explored.

3.5 ROS Signal Communication During Cell Death

ROS-dependent apoptotic cell death involving mitochondria has long been considered as fundamental mechanism in animal cell. The discharge of cytochrome c essential for the activation of caspases paved the mode of action of nuclear condensation, depolarization of mitochondrial membrane, and other symptoms associated with animal cell apoptosis (Fleury et al. 2002; Doke 1997). When compared with plants, though chloroplast takes pivotal role, mitochondria may work as a major site of ROS formation, where the early changes in ROS homeostasis is augmented, activating the release of cytochrome c through the opening of mitochondrial transition pore (Dat et al. 2003; Casolo et al. 2005; Torres and Dangl 2005). The function of mitochondria in ROS signaling during plant apoptosis is further corroborated by the upregulation of activities of antioxidative defense enzymes like Mn-SOD, alternative oxidase, etc. (Dat et al. 2003). Chloroplasts in plant cell also are the site of origin of ROS signal that subsequently generates retrograde signals involved in plant PCD. For example, during cryptogein-induced PCD, the ROS burst, with H_2O_2 as the prime component, induces upregulation in the activity of lipoxygenases (LOX). Activated LOX, in turn, targets chloroplastic PUFAs, having cis,cis, 1,4-pentadiene moiety, causing membrane lipid peroxidation (MLP), producing oxylipins which finally instigates PCD (Maccarrone et al. 2000; Montillet et al. 2005) (Fig. 3.2). The ectopic expression of mammalian BCL2 members in the chloroplasts, which protect transgenic tobacco plants from herbicide-induced PCD, proves the chloroplastic involvement in PCD (Chen and Dickman 2004). Further, the role of phytochrome signaling during the initiation of HR clearly substantiates the necessity of chloroplastic factor(s) in the mechanism associated with the HR (Genoud et al. 2002; Karpinski et al. 2003). Thus, apart from the well-documented contribution of mitochondria in PCD, ample of evidences support the role of chloroplast-derived signals during plant PCD.

Involvement of other signaling molecules (jasmonate, salicylic acid, ethylene) along with ROS, through their feed forward or backward interactions, exists during plant PCD (Overmyer et al. 2005). ROS-dependent PCD is connected with enhanced levels of both salicylic acid (SA) and ethylene. The fact that the activation and upregulation of SA-degrading enzymes in ethylene mutants has been found to delay PCD, strongly conveys the positions of both the growth regulators within a signaling circuits that promote ROS-dependent cell death (Moeder et al. 2002; Danon et al. 2005). However, the growth regulator jasmonic acid can either upregulate or downregulate in this ROS-dependent PCD, depending on the type of ROS generated. Further, different interaction with ROS-instigated MLPO products or oxylipins adds a further level of specificity and intricacy to ROS signaling during PCD (Danon et al. 2005) (Fig. 3.2).

Molecular studies further substantiate the ROS-dependent pathway in plants. Gene isolated through a genetic screen in *Arabidopsis* lsd1 corroborates our understanding of ROS-dependent PCD. The lsd1 mutant of *Arabidopsis* fails to survive under elevated level of ROS superoxide and exhibited natural superoxide- and SA-dependent runaway necrotic lesions (Jabs et al. 1996). The functional role of LSD1 is to encode a zinc-finger protein, which along with two other zinc-finger proteins (LOL1and LOL2) helps in sensing the changes in redox status, by working as a molecular rheostat, thereby downregulating senescence or regulating an anti-senescence mechanism by controlling the expression of cell death genes. There are also instances, where MAPKs found to be involved downstream in ROS-dependent senescence. In tobacco, the ROS-activated MAPK is basically a protein kinase, also induced by salicylic acid, which is involved in harpin-dependent PCD (Samuel et al. 2005). In another instance of *Medicago sativa*, a MAPKKK activates cell death pathway after getting oxidatively modified by ROS through a precise MAPK-scaffolding mechanism (Nakagami et al. 2004). Though there are some homology between plant and animal cell death mechanism, no clear homologs of the BCL-like animal cell death suppressors have been identified in plant genomes till date. Further, the protection against ROS-dependent cell death, originated in chloroplast and mitochondria, is bestowed by their excessive in tobacco. Though there are evidences of functional homologs in defense of ROS-mediated PCD in transgenic plants, their presence under normal system still remains uncertain (Mitsuhara et al. 1999; Chen and Dickman 2004). However, Kang et al. (2006) identified a BCL2-associated protein, which happen to be evolutionarily conserved in *Arabidopsis* and also found to be upregulated by ROS H_2O_2 and capable of instigating PCD in both yeast and plants.

3.6 Conclusion and Perspective

There is ample evidence to link redox status of the cell with plant senescence. Understanding the relationship between senescence and redox status of the cell not only provides regulatory role of ROS on senescence but also offers new tools for the improvement of agricultural crops. Current researches on redox-dependent PCD has demonstrated that loss of redox homeostasis of the cell leads to significant accumulation of ROS, which might be a key factor for dismantling of cellular organelle and cell death. ROS accumulation beyond threshold level became phytotoxic as they indiscriminately cause oxidative damage to all major cellular macromolecules, causing oxidative deterioration of cell. The ROS-dependent necrosis and is commonly encountered under cellular conditions with serious loss of redox homeostasis due to leakage of electrons from ETCs having reduced antioxidant potential. Environmental stresses may also enhance the probability of cell death being steered through ROS over-accumulation and associated redox imbalance. In certain cases, when the accumulation of ROS due to extrinsic factors (stresses) is not sufficient enough to cause cell death directly, subtle changes in internal redox cue appear to initiate a signaling event, leading to PCD, as typically happened during HR. The identification and characterization of plant genes associated with ROS-dependent PCD are just the beginning of a new era for unfolding molecular basis of senescence and its redox regulation.

Though apparently there seems to have similarity between plant and animal senescence, till now, no functional homologs of known apoptotic regulators, such as caspases or BCL2, have been detected in plants. However, mutant screening helped us to identify suppressor of ROS-induced cell death. Cloning of such alleles is reported to revert or induce PCD in *Arabidopsis* mutants and helped in the identification of such genes in the plant PCD pathway schemes. Comparative transcriptomic studies revealed several genes associated with PCD which are induced by ROS signals (Desikan et al. 2001; op den Camp et al. 2003; Vandenabeele et al. 2003; Vanderauwera et al. 2005; Gadjev et al. 2006). Genes involved in ROS perception and signal transduction associated with PCD are yet to be discovered. Practical screens, in which these categories of genes are differentially regulated and scored for their prospective role to cause ROS-dependent cell death, will definitely assist in their detection. So, this approach, together with the characterization of redox sensitive target proteins that undergo posttranslational modifications, such as protein oxidation and nitrosylation, followed by their regulation during ROS-dependent cell death pathway, will help us to unfold the molecular mechanism associated with the event.

References

Apel K, Hirt H (2004) Reactive oxygen species: metabolism, oxidative stress, and signal transduction. Annu Rev Plant Biol 55:373–399

Arora A, Sairam RK, Srivastava GC (2002) Oxidative stress and antioxidative system in plants. Curr Sci 82:1227–1238

Asada K (1994) Production and action of active oxygen in photosynthetic tissue. In: Foyer CH, Mullineaux PM (eds) Causes of photooxidative stress and amelioration of defense systems in plants. CRC Press, Boca Raton, pp 77–104

Beckman KB, Ames BN (1997) Oxidative decay of DNA. J Biol Chem 272:19633–19636

Bhattacharjee S (2005) Reactive oxygen species and oxidative burst: roles in stress, senescence and signal transduction in plants. Curr Sci 89:113–1121

Bhattacharya K, Naha PC, Naydenova I, Mintova S, Byrne HJ (2012) Reactive oxygen species mediated DNA damage in human lung alveolar epithelial (A549) cells from exposure to non-cytotoxic MFI-type zeolite nanoparticles. Toxicol Lett 215(3):151–160

Casolo V, Petrussa E, Krajnˇ a´kova´ J, Macrì F, Vianello A (2005) Involvement of the mitochondrial KTP channel in H2O2- or NO-induced programmed death of soybean suspension cell cultures. J Exp Bot 56:997–1006

Chen S, Dickman MB (2004) Bcl-2 family members localize to tobacco chloroplasts and inhibit programmed cell death induced by chloroplast targeted herbicides. J Exp Bot 55:2617–2623

Dangl JL, Dietrich RA, Thomas H (2000) Senescence and programmed cell death. In: Buchanan B, Gruissem W, Jones R (eds) Biochemistry and molecular biology of plants. American Society of Plant Physiologists Press, Rockville, pp 1044–1100

Danon A, Miersch O, Felix G, op den Camp RGL, Apel K (2005) Concurrent activation of cell death-regulating signaling pathways by singlet oxygen in *Arabidopsis thaliana*. Plant J 41:68–80

Dat JF, Pellinen R, Beeckman T, Van De Cotte B, Langebartels C, Kangasjärvi J, Inzé D, Van Breusegem F (2003) Changes in hydrogen peroxide homeostasis trigger an active cell death process in tobacco. Plant J 33:621–632

Desikan R, Mackerness SA-H, Hancock JT, Neill SJ (2001) Regulation of the Arabidopsis transcriptome by oxidative stress. Plant Physiol 127:159–172

Doke N (1997) The oxidative burst: roles in signal transduction and plant stress. In: Scandalios JG (ed) Oxidative stress and the molecular biology of antioxidant defenses, Monograph series, vol 34. Cold Spring Harbor Laboratory Press, Cold Spring Harbor, pp 785–813

Drew MC, He C-J, Morgan PW (2000) Programmed cell death and aerenchyma formation in roots. Trends Plant Sci 5:123–127

Fleury C, Mignotte B, Vayssière J-L (2002) Mitochondrial reactive oxygen species in cell death signaling. Biochimie 84:131–141

Foyer CH, Noctor G (2005) Redox homeostasis and antioxidant signaling: a metabolic interface between stress perception and physiological responses. Plant Cell 17:1866–1875

Fridovich I (1995) Superoxide radical and SODs. Annu Rev Biochem 64:97–112

Gadjev I, Vanderauwera S, Gechev TS, Laloi C, Minkov IN, Shulaev V, Apel K, Inzé D, Mittler R, Van Breusegem F (2006) Transcriptomic footprints disclose specificity of reactive oxygen species signaling in Arabidopsis. Plant Physiol 141:436–445

Gan S, Amasino RM (1997) Making sense of senescence (molecular genetic regulation and manipulation of leaf senescence). Plant Physiol 113:313–319

Genoud T, Buchala AJ, Chua N-H, Métraux J-P (2002) Phytochrome signalling modulates the SA-perceptive pathway in Arabidopsis. Plant J 31:87–95

Gunawardena AHLAN, Greenwood JS, Dengler NG (2004) Programmed cell death remodels lace plant leaf shape during development. Plant Cell 16:60–73

Jabs T (1999) Reactive oxygen intermediates as mediators of programmed cell death in plants and animals. Biochem Pharmacol 57:231–245

Jabs T, Dietrich RA, Dangl JL (1996) Initiation of runaway cell death in an Arabidopsis mutant by extracellular superoxide. Science 273:1853–1856

Jacobson MD, Weil M, Raff MC (1997) Programmed cell death in animal development. Cell 88:347–354

Jajic I, Sarna T, Szewczyk G, Strzalka K (2015) Changes in production of reactive oxygen species level in illuminated thylakoids isolated during development and senescence of barley leaf senescence. J Plant Physiol 184:49–56

Kang CH, Jung WY, Kang YH, Kim JY, Kim DG, Jeong JC, Baek DW, Jin JB, Lee JY, Kim MO (2006) AtBAG6, a novel calmodulin-binding protein, induces programmed cell death in yeast and plants. Cell Death Differ 13:84–95

Karpinski S, Gabrys H, Mateo A, Karpinska B, Mullineaux PM (2003) Light perception in plant disease defence signalling. Curr Opin Plant Biol 6:390–396

Khanna-Chopra R, Selote DS (2007) Acclimation to drought stress generates oxidative stress tolerance in drought-resistant than -susceptible wheat cultivar under field conditions. Environ Exp Bot 60:276–283. https://doi.org/10.1016/j.envexpbot.2006.11.004

Krishnamoorthy RR, Crawford MJ, Chaturvedi MM, Jain SK, Aggarwal BB, Al-Ubaidi MR, Agarwal N (1999) Photo-oxidative stress down-modulates the activity of nuclear factor-kappaB via involvement of caspase-I, leading to apoptosis of photoreceptor cells. J Biol Chem 274:3734–3743

Kuo A, Cappelluti S, Cervantes-Cervantes M, Rodriguez M, Bush DS (1996) Okadaic acid, a protein phosphatase inhibitor, blocks calcium changes, gene expression, and cell death induced by gibberellin in wheat aleurone cells. Plant Cell 8:259–269

Kuriyama H, Fukuda H (2002) Developmental programmed cell death in plants. Curr Opin Plant Biol 5:568–573

Lam E (2004) Controlled cell death, plant survival and development. Nat Rev Mol Cell Biol 5:305–315

Langebartels C, Wohlgemuth H, Kschieschan S, Grün S, Sandermann H (2002) Oxidative burst and cell death in ozone-exposed plants. Plant Physiol Biochem 40:567–575

Lee S, Seo PJ, Lee HJ, Park CM (2012) A nac transcription factor ntl4 promotes reactive oxygen species production during drought-induced leaf senescence in arabidopsis. Plant J Cell Mol Biol 70:831–844. https://doi.org/10.1111/j.1365-313X.2012.04932.x

Levine A, Tenhaken R, Dixon R, Lamb C (1994) H2O2 from the oxidative burst orchestrates the plant hypersensitive disease resistance response. Cell 79:583–593

Lorrain S, Vailleau F, Balagué C, Roby D (2003) Lesion mimic mutants: keys for deciphering cell death and defense pathways in plants? Trends Plant Sci 8:263–271

Lushchak VI (2011) Adaptive response to oxidative stress: bacteria, fungi, plants and animals. Comp Biochem Physiol Toxicol Pharmacol CBP 153:175–190. https://doi.org/10.1016/j.cbpc.2010.10.004

Maccarrone M, Van Zadelhoff G, Veldink GA, Vliegenthart JFG, Finazzi- Agrò A (2000) Early activation of lipoxygenase in lentil (*Lens culinaris*) root protoplasts by oxidative stress induces programmed cell death. Eur J Biochem 267:5078–5084

Mitsuhara I, Malik KA, Miura M, Ohashi Y (1999) Animal cell-death suppressors Bcl-xL and Ced-9 inhibit cell death in tobacco cells. Curr Biol 9:775–778

Mittler R, Hallak Herr E, Orvar BL, Van Camp W, Willekens H, Inzé D, Ellis BE (1999) Transgenic tobacco plants with reduced capability to detoxify reactive oxygen intermediates are hyperresponsive to pathogen infection. Proc Natl Acad Sci 96:14165–14170

Mittler R, Vanderauwera S, Gollery M, Van Breusegem F (2004) The reactive oxygen gene network in plants. Trends Plant Sci 9:490–498

Moeder W, Barry CS, Tauriainen AA, Betz C, Tuomainen J, Utriainen M, Grierson D, Sandermann H, Langebartels C, Kangasjärvi J (2002) Ethylene synthesis regulated by biphasic induction of 1-aminocyclopropane-1-carboxylic acid synthase and 1-aminocyclopropane-1-carboxylic acid oxidase genes is required for hydrogen peroxide accumulation and cell death in ozone-exposed tomato. Plant Physiol 130:1918–1926

Montillet J-L, Chamnongpol S, Rustérucci C, Dat J, Van de Cotte B, Agnel J-P, Battesti C, Inzé D, Van Breusegem F, Triantaphylidès C (2005) Fatty acid hydroperoxides and H2O2 in the execution of hypersensitive cell death in tobacco leaves. Plant Physiol 138:1516–1526

Nakagami H, Kiegerl S, Hirt H (2004) OMTK1, a novel MAPKKK, channels oxidative stress signaling through direct MAPK interaction. J Biol Chem 279:26959–26966

Op den Camp RGL, Przybyla D, Ochsenbein C, Laloi C, Kim C, Danon A, Wagner D, Hideg E, Göbel C, Feussner I (2003) Rapid induction of distinct stress responses after the release of singlet oxygen in Arabidopsis. Plant Cell 15:2320–2332

Overmyer K, Brosché M, Pellinen R, Kuittinen T, Tuominen H, Ahlfors R, Keinänen M, Saarma M, Scheel D, Kangasjärvi J (2005) Ozone-induced programmed cell death in the Arabidopsis *radical-induced cell death1* mutant. Plant Physiol 137:1092–1104

Pennell RI, Lamb C (1997) Programmed cell death in plants. Plant Cell 9:1157–1168

Polle A (2001) Dissecting the superoxide dismutase-ascorbate-glutathione pathway in chloroplasts by metabolic modeling: computer simulations as a step towards flux analysis. Plant Physiol 126:445–462

Samuel MA, Hall H, Krzymowska M, Drzewiecka K, Hennig J, Ellis BE (2005) SIPK signaling controls multiple components of harpin-induced cell death in tobacco. Plant J 42:406–416

Selote D, Khanna-Chopra R (2006) Drought acclimation confers oxidative stress tolerance by inducing co-ordinated antioxidant defense at cellular and subcellular level in leaves of wheat seedlings. Physiol Plant 127:494–506. https://doi.org/10.1111/j.1399-3054.2006.00678.x

Souter M, Lindsey K (2000) Polarity and signalling in plant embryogenesis. J Exp Bot 51:971–983

Thompson JE, Legge RL, Barber RF (1987) The role of free radicals in senescence and wounding. New Phytol 105:317–344. https://doi.org/10.1111/j.1469-8137.1987.tb00871.x

Thompson JE, Froese CD, Madey E, Smith MD, Hong Y (1998) Lipid metabolism during plant senescence. Prog Lipid Res 37:119–141

Torres MA, Dangl JL (2005) Functions of the respiratory burst oxidase in biotic interactions, abiotic stress and development. Curr Opin Plant Biol 8:397–403

Vacca RA, de Pinto MC, Valenti D, Passarella S, Marra E, De Gara L (2004) Production of reactive oxygen species, alteration of cytosolic ascorbate peroxidase, and impairment of mitochondrial metabolism are early events in heat shock-induced programmed cell death in tobacco Bright-Yellow 2 cells. Plant Physiol 134:1100–1112

Van Doorn WG, Woltering EJ (2005) Many ways to exit? Cell death categories in plants. Trends Plant Sci 10:117–122

Vandenabeele S, Van Der Kelen K, Dat J, Gadjev I, Boonefaes T, Morsa S, Rottiers P, Slooten L, Van Montagu M, Zabeau M (2003) A comprehensive analysis of hydrogen peroxide-induced gene expression in tobacco. Proc Natl Acad Sci U S A 100:16113–16118

Vanderauwera S, Zimmermann P, Rombauts S, Vandenabeele S, Langebartels C, Gruissem W, Inzé D, Van Breusegem F (2005) Genome-wide analysis of hydrogen peroxide-regulated gene expression in Arabidopsis reveals a high light-induced transcriptional cluster involved in anthocyanin biosynthesis. Plant Physiol 139:806–821

Vranová E, Inzé D, Van Breusegem F (2002) Signal transduction during oxidative stress. J Exp Bot 53:1227–1236

Wang L, Su H, Han L, Wang C, Sun Y, Liu F (2014) Differential expression profiles of poplar map kinase kinases in response to abiotic stresses and plant hormones, and overexpression of PtMKK4 improves the drought tolerance of poplar. Gene 545:141–148. https://doi.org/10.1016/j.gene.2014.04.058

Young TE, Gallie DR (2000) Programmed cell death during endosperm development. Plant Mol Biol 44:283–301

ROS and Oxidative Modification of Cellular Components

4

Abstract

ROS, the inevitable by-products of aerobic metabolism, when got escaped from antioxidant-mediated detoxification and accumulated in high concentration, may react nonspecifically with almost all important biomolecules and cause irreversible damage to those biomolecules which may cause metabolic dysfunction and inactivation of key cellular functions. In fact, there exist several evidences on environmental stress (both abiotic and biotic)-mediated changes in redox status and corresponding modulation of lipid and protein oxidation. ROS-mediated peroxidation of lipid, particularly the membrane lipid peroxidation (MLPO), which is normally linked with aging, senescence, and stress-induced oxidative damages, is extremely important from its mechanistic point of view as it is a unique example of hydrocarbon-centered ROS production in cell. Oxidative modification of proteins, which basically involves through carbonylation, nitrosylation, disulfide linkage formation, glutathionylation, etc., seriously modifies and influences the protein activity and subsequently causes metabolic dysfunction. Though oxidative modifications of important cellular molecules exhibit deteriorative events, in recent times, several works highlighted important physiological roles of such modifications. The most important of which is membrane lipid peroxidation (MLPO). Mechanistically it involves the generation and propagation of lipid radicals, oxygenation and rearrangement of double bonds in the unsaturated lipids, and in due course their destruction. This process always yields a variety of secondary breakdown products, including alcohol, aldehydes, ketones, alkanes, and ethers, and is long being recognized as one of the prime events involved in oxidative damage and cell death. However, several recent works propose that these secondary products of lipid peroxidation, like reactive lipid species (RLS), can instigate cell signaling associated with the induction of antioxidative defense, apoptosis, membrane repair, etc. In this chapter an effort has been made to provide an update on oxidative damages of important cellular components while addressing their roles in deteriorative oxidative damage and adaptive cell signaling. This chapter makes an effort to examine the biochemical

S. Bhattacharjee, *Reactive Oxygen Species in Plant Biology*,
https://doi.org/10.1007/978-81-322-3941-3_4

aspects of oxidative damage to lipid, protein, and nucleic acids along with the technologies to detect those changes. An additional effort has also been made to unfold the physiological significance of oxidative damages of lipid and protein in context of the recent roles of secondary oxidized products in cell signaling.

Keywords
Reactive oxygen species · Reactive lipid species · Lipid peroxidation · Lipid peroxidation products · Protein oxidation · Cell signaling

4.1 Introduction

Unfavorable environmental conditions such as drought, salinity, extremes of temperature, radiation, excess photochemical energy, etc. have serious impact on plant metabolism and productivity (Cramer et al. 2011; Gill and Tuteja 2010; Krasenky and Jonak 2012). Any change in environmental parameters or even infection causes changes in redox status or homeostasis of the cell toward accretion of prooxidants causing origin of oxidative stress. Antioxidative defense system comprising of enzymatic and nonenzymatic defense system works systematically to regulate the endogenous titer of prooxidants and keep their titer to nontoxic levels (Mittler 2002; Gill and Tuteja 2010; Bhattacharjee 2005, 2012a, b). However, incapability of the plant to offer adequate antioxidative security by producing sufficient and different capability makes them fall prey to abiotic stress-induced oxidative stress and subsequent damages. Stresses either alone or in combination cause severe damage to plants by triggering oxidation of important metabolites under oxidative stress. Although there are evidences of tight regulation of endogenous titer ROS necessary for inducing cell signaling under environmental stress, any disturbance in redox homeostasis in cell always leads to oxidative threat for biomolecules. This loss of redox homeostasis or oxidative stress always leads to significant physiological challenges including adverse changes in plant metabolism, growth, and development and cell death. The oxidative threat under oxidative stress mainly includes oxidative changes of essential biomolecules including lipids, proteins, amino acids, DNA, etc. (Krasenky and Jonak 2012; Gill and Tuteja 2010), which eventually has several metabolic, molecular, and physiological consequences, culminating in loss of yield and productivity of crops (Fig. 4.1).

Under oxidative threat, peroxidation of lipid and oxidation of protein are the most extensively studied consequences of ROS action which have tremendous effect on architecture and role of the membrane (Foyer and Shigeoka 2011; Davies 2005). Enhanced MLPO and protein oxidation (PO), which are symptoms of stress-induced injury, may take place in the membrane of cell and organelles that ultimately affect several metabolic homeostasis of the cell. Further, peroxidation of membrane lipid can aggravate oxidative stress by generating secondary oxidative products like alkoxy radicals, peroxy radicals, singlet oxygen, etc. which eventually can damage proteins and DNA (Sharma et al. 2012). The ROS-mediated protein

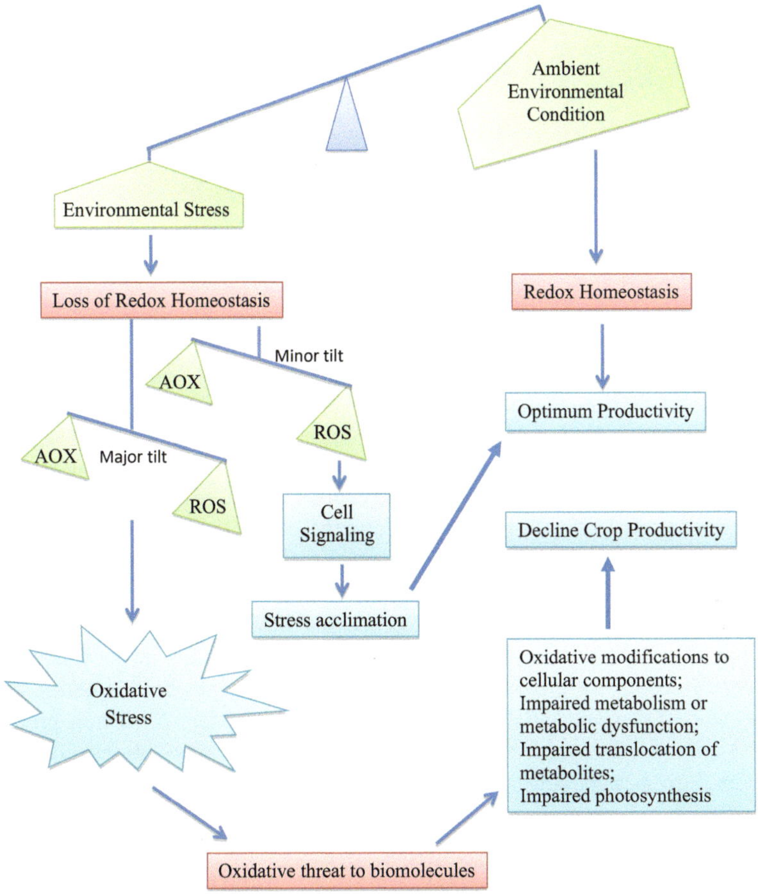

Fig. 4.1 Schematic representation of the impact of changes in redox homeostasis triggered by environmental stress on oxidative modification, and subsequent impact on crop productivity

oxidation may be site specific in nature which eventually causes fragmentation of peptide chains, aggravation of cross-linked secondary products, and enhanced susceptibility for ubiquitylation. Nevertheless, the MLPO and PO are found to be linked and may take place in tandem, where the lipid-peroxidized products can interact with proteins (Yamauchi et al. 2008). In fact, the extremely short half-life of ROS compelled the workers to standardize procedures for qualifying ROS-based oxidize products of lipid and proteins as an alternative approach for assessing the stress-induced cytotoxicities in plant. In this context the estimation of TBARS and free carbonyl content (RC=O) has been used as biochemical marker for oxidative stress (Chakraborty and Bhattacharjee 2015; Sochor et al. 2012).

Based on this background, this chapter examines the chemistry and physiological implications of oxidative damages to protein, lipid, and nucleic acids. The present study also made an inventory of recent technologies used to evaluate the oxidative deterioration of protein and lipids. Special emphasis was green given on the cross-links between protein lipid oxidation and the role of secondary oxidized products in cell signaling.

4.2 Oxidative Modification of Membrane Lipid

Oxidative modification of membrane fatty acids through peroxidation, particularly the membrane-associated polyunsaturated fatty acids (PUFAs) that are sensitive to oxidation, is recognized as the Most important mechanisms involved in oxidative membrane deterioration and eventually the cell architecture which may lead to cell death (Dianzani and Barrera 2008; Farooqui and Farooqui 2011; Skorzynska-Polit 2007; Bhattacharjee 2005). The loss of redox homeostasis under environmental stress, along with the abundance of membrane-bound PUFAs and presence of transition metal ions in cellular atmosphere, may be considered as ideal situation for initiating LPO (Foyer and Noctor 2013; Bhattacharjee 2012a, b; Halliwell and Gutteridge 1999; Aust et al. 1995). In fact, the occurrence of transition metal ions along with high concentration of lipid-hydrolyzed product PUFAs (having certain structural identity), generated significantly under different kinds of environmental stress, perfectly sets the condition to initiate the process of lipid peroxidation. MLPO in plant cell may be enzymatic and nonenzymatic and always results in the formation of a number of secondary breakdown products (aldehydes, alcohols, ketones, alkanes, and ethers) (Dianzani and Barrera 2008). These secondary oxidation products of MLPO are often recognized as biomarkers of oxidative stress associated with environmental stresses and senescence (Higden et al. 2012; Pralico et al. 2001; Thompson et al. 1987). The products of MLPO (detected by different analytical techniques) often show positive correlation with stress-induced metabolic dysfunction and physiological impairment (Morrow 2005; Marwah et al. 2002). A different concept of membrane lipid peroxidation emerges in recent times with the identification of reactive lipid species (RLS), a class of breakdown products of oxidized lipids that are electrophilic in nature and are competent enough to interact with cellular nucleophiles such as amino acids, peptides, etc. it is suggested that this group of radical species (RLS) are not simply the by-products of MLPO but are the intermediates in MLPO pathway under multiple physiological conditions. Most importantly, RLS, being electrophilic, can form stable covalent adducts with nucleophilic residues on proteins (Higden et al. 2012; Landar et al. 2006; Codreanu et al. 2009). This mechanism of RLS reaction to cellular nucleophiles is extremely significant and vital as the redox-sensitive thiol – Switches of peptides and specific amino acids control a plethora of cell signaling events or even differentially regulate gene expressions (Higden et al. 2012; Rudolph and Freeman 2009). In fact, the thiol-reactive signaling intermediates of amino acids and peptides seems to be biologically modulated through their interaction with RLS. Thus oxidative modification-mediated ROS or secondary breakdown products of MLPO ultimately to compounds such as sulfinic or sulfenic acids, which were at the outset recognized as manifestation of oxidative stress, are now suggestive of their role in cell acclamatory signaling. So, the concept and physiological implication of MLPO are gradually shifting, and a judicious view of the process of lipid peroxidation is coming up, suggesting the fact that apart from its deteriorative aspect, MLPO products can elicit diverse cellular effects depending on their reactivity, abundance, types, and target molecules.

4.2.1 Mechanism of Membrane Lipid Peroxidation (MLPO)

Peroxidation of fatty acids of membrane lipids is important from its mechanistic point of view due its ability to generate free radicals, and hence it is considered as one of the few examples of carbon-centered oxyfree radical or ROS production in plant cell (Winston 1990). MLPO essentially involves four distinct stages, namely, initiation, progression, termination, and propagation. The mechanism involves cascading chain reactions which is at the outset triggered by abstraction of hydrogen atom (refer to Fig. 1.5, Reaction 1) or an addition of oxygen radical that causes oxidative breakdown of membrane-bound or membrane-derived PUFA. PUFAs being more susceptible to oxidation as compared to saturated fatty acids, the activated methylene bridge (-LH) became a critical oxidative target site. In the next step of propagation, molecular oxygen rapidly adds on, i.e., oxygenation to carbon-centered radical ($L^•$) takes place, forming peroxy radical ($LOO^•$) (refer to Fig. 1.5, Reactions 2–5). Subsequently, in propagation step, $LOO^•$ generates organic hydroperoxides, which in turn can abstract H from a new PUFA, initiating the process once again (Fig. 4.2). Subsequently, the alkyl radicals may be stabilized by rearrangement into fatty acid dimer or conjugated diene (Fig. 1.5), which are relatively stable products of MLPO. Therefore, lipid hydroperoxides are the first stable product of MLPO, and the others are conjugated diene. However, in other cases, where MLPO is continuously spiraling, particularly under severe redox imbalance, radical termination might occur with the formation of peroxy-bridged dimmers (Fig. 1.5, Reactions 6–8).

The transition metal ions therefore initiate the event of MLPO. These metal complexes either form activated oxygen complexes which can abstract allelic hydrogen or even can work as a catalyst in the breakdown of existing lipid hydroperoxides. Though ROS like $O_2^{•-}$ and H_2O_2 are capable to initiate MLPO, $OH^•$ being the most potent one, the initiation of MLPO is mainly mediated by $OH^•$. Sometimes, the loosely bound Fe (Fe-containing peptides) is also able to initiate the breakdown of lipid hydroperoxides causing the accumulation of alkoxy and peroxy radicals, which may further stimulate propagation event of MLPO. The physical structures of plant membranes also facilitate autocatalytic propagation of lipid peroxidation by placing the fatty acid side chains in close proximity. Fe being a transition metal that can exist in several valences (which can be bound up to six ligands) is an important component for redox reactions of MLPO (Boveris et al. 2008). Several other facts also corroborate strongly the role of transition metal in MLPO. For example, some transition metal ions like Fe, Pb, Cd, and Cr can induce MLPO effect under in vitro condition (Repetto and Boveris 2012; Repetto 2008). However the pace of the Fenton reaction is very low and hence fails to contribute substantial formations of $OH^•$ necessary for

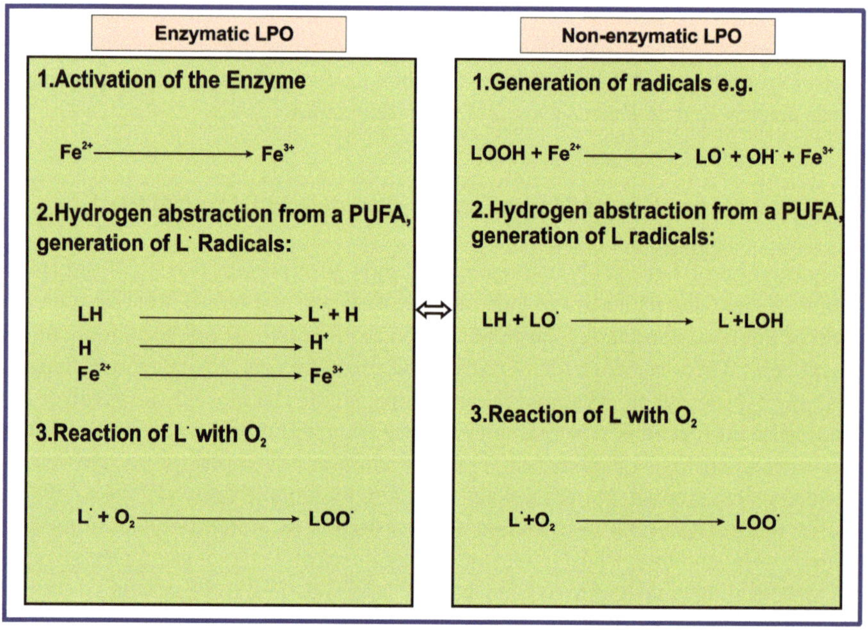

Fig. 4.2 Events associated with enzymatic and nonenzymatic MLPO. (Detail in text)

initiation of MLPO in cellular system. However, the nanomolar concentration of free transition metal elements or their loosely associated forms seems adequate for sponsoring Fenton reaction in vitro at physiological level of ROS (Repetto et al. 2010).

4.2.2 Enzymatic and Nonenzymatic Membrane Lipid Peroxidation (MLPO) Works in Tandem in Plant Cell with Physiological Relevance

Mechanistically, the MLPO in plant cell may be enzymatic or nonenzymatic (Fig. 4.2). The enzyme lipoxygenase mediates enzymatic MLPO, whereas ROS itself are found to stimulate nonenzymatic MLPO. This enzyme is able to initiate the formation of fatty acid hydroperoxides, ensuing peroxidation (Spiteller and Spiteller 1997). The enzyme lipoxygenase (LOX) is found to be activated or upregulated particularly under stress and senescence, targeting specific PUFAs. PUFAs which are characterized by the presence of -CH=CH-CH_2-CH=CH- in its skeleton (cis,cis-1,4- penta-diene moiety) became the target of LOX and transform them into lipid hydroperoxides (LOOHs) (Bhattacharjee 2014). The hydroperoxides (LOOH) being unstable are decomposed to a great variety of products. LOX are also capable enough to cleave in a regio- and stereospecifically controlled reaction an H atom from a double allylically activated –CH_2 group of PUFA. While still remaining attached to LOX, the H atom reacts with complex attached with Fe^{3+} in the active form of LOX and forms H^+ and Fe^{2+} ion. Lipid radicals (L•) formed subsequently add oxygen and produce peroxyl radical (LOO•). Eventually, an electron migrates from Fe^{2+} LOO•, producing peroxyl anion (LOO^-), which combines with H^+ to produce LOOH (deGroot et al. 1975). It is important to note that during the entire process of LOX-mediated MLPO, the peroxyl radical remains attached with the enzyme complex and not being able to escape (Jabs 1999).

In plants, several isoforms of the enzyme LOX have been detected which can be distinguished by variations in their kinetic properties and localization and developmental stages of their activity (Fuller et al. 2001). The physiological role of LOX-mediated MLPO ranges from mobilization of storage lipids fatty seeds during germination to the formation of ROS under unfavorable environmental condition and senescence (Feussner et al. 1997; Porta and Rocha-Sosa 2002). The close relationship between senescence and MLPO can be substantiated by an accumulation of MLPO products and ROS along with aging (Jabs 1999; Spreitzer et al. 1989; Feussner and Wasternack 2004).

Nonenzymatic membrane lipid peroxidation (NE-MLPO), which is nonspecific in nature, also proceeds through a similar chain reaction exhibiting initiation, progression, termination, and propagation events as already discussed. However, in E-MLPO, initiation by stereospecific propagation does not occur frequently (Higden et al. 2012). In case of NE-MLPO, the allelic H atom in PUFA is abstracted by ROS and results in the formation of lipid radicals, which subsequently react with O_2. Different products, depending on the nature of fatty acid oxidized, are generated. Once NE-MLPO got initiated, both the lipid alkoxyl radical (LO•) and lipid peroxyl radical (LOO•) capable of abstracting H from another fatty acid molecule are formed, thus reinitiating the propagation of MLPO (Fig. 4.2). Membrane proteins also assist in MLPO by stimulating propagation event of MLPO. Lipid radicals can be transferred to protein side chains with subsequent adduct formation, thereby causing the proteins to be an active participants in the propagation of MLPO. Finally, the nonenzymatic lipid peroxidation terminates into radical-radical reactions (Fritz and Petersen 2011).

Experimental evidences support the fact that LOX-mediated MLPO often switches over to a nonenzymatic MLPO, when supply of substrate for MLPO (PUFA) exceeds certain threshold limit (Spiteller 2003; Fuchs and Spiteller 2000). In this case, a tandem action of both enzymatic and nonenzymatic MLPO was observed, as in case of apoptotic event. Paradoxically it is found that, when the availability of PUFA is significantly elevated, LOX not only catalyze MLPO but also commit suicide by catalyzing degradation of its own enzyme molecule. This eventually causes the release of enzyme-bound Fe-ion, which subsequently reacts with end product of LOX-mediated MLPO (i.e., LOOH) to produce LO• in a Fenton-type reaction (Spiteller 2003). Lipid radicals formed in this manner then abstract a fresh H atom from double allylic activated CH_2 group of another PUFA, forming a new lipid radical (L•), thus inducing once again the chain reaction (Fig. 4.2). The alkoxyl radical is produced in the event primarily by reaction of LOOH with Fe^{2+}, although most of the LOOH are reduced by peroxidases to corresponding OH• (Wang and Powell 1991). Physiologically, the alkoxyl radicals generated are subsequently decomposed to 2,4-unsaturated aldehydes (mainly

2,4-dienols) and alkyl radicals (R$^\bullet$). The 2,4-dienal may be further converted to other secondary aldehyde compounds (Fig. 4.3).

In a study with simplest PUFA, linoleic acid (generally a preferred substrate for LOX-induced MLPO), transformation during the event causes formation of 9-hydroxyperoxy-10, 12-octadecadienoic acid (9-HPODE) and 13-hydroxyperoxy -9, 11-decadinoic acid (13-HPODE). HPODEs are subsequently reduced to their corresponding alcohols. These classes of compounds represent the major membrane lipid peroxidation products that accumulate due to tandem action of both enzymatic and nonenzymatic lipid peroxidation in drought stressed and senescing plant tissues.

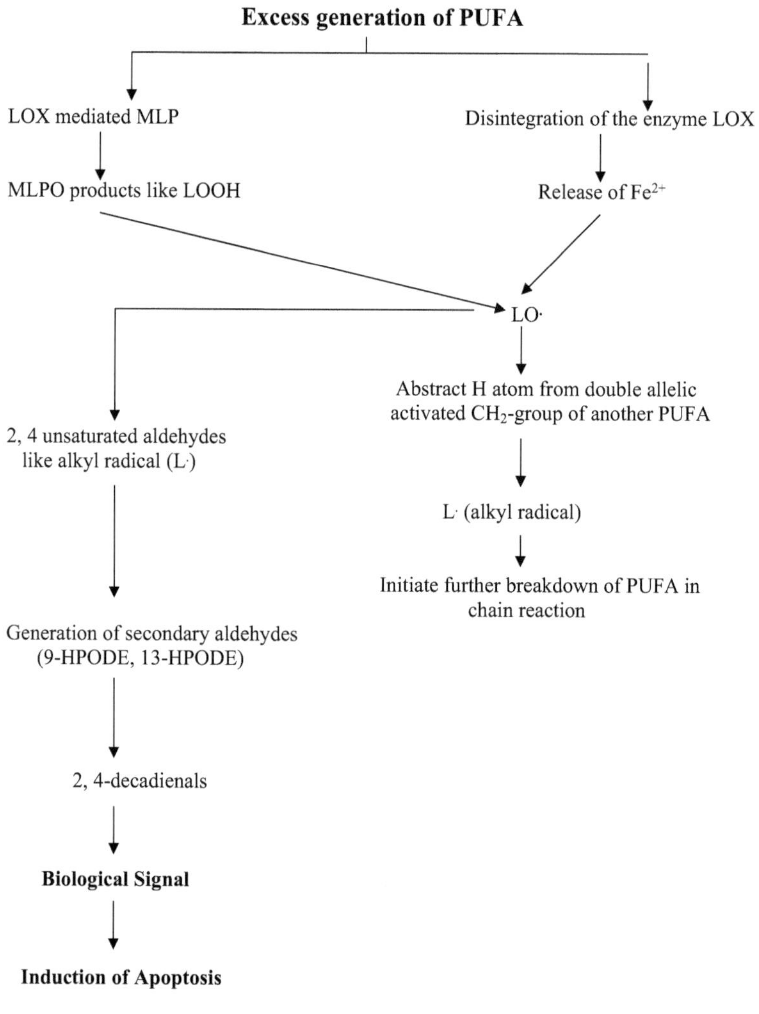

Fig. 4.3 Implication of MLPO in the origin of "Biological Signal." (Detail in text)

4.2.3 MLPO Serves Other Physiological Purpose Apart from Oxidative Damage

MLPO, particularly the nonenzymatic type (catalyzed by ROS itself), is viewed as deleterious one, but wide-ranging study of this process in plant in the last decade proved other important cellular functions of this process as well (Skorzynska-Polit 2007; Foyer and Noctor 2013; Bhattacharjee 2012a, b). The classical view of environmental stress-induced oxidative stress-mediated MLPO that erratically destroys cell membranes has undergone a paradigm shift, in which several positive biological roles of this process are becoming gradually evident. The membrane-associated O_2-sensitive PUFAs when get peroxidized under a condition of redox imbalance produce several secondary products (RLS), whose biological roles in stimulating expression of genes encoding detoxification or defense functions or the repair of damaged membrane seem to be extremely important.

Among the targets of ROS, under a situation of loss of redox homeostasis influenced by abiotic and biotic stresses, decontrolled MLPO is considered as most damaging event. It not only generates a host of ROS and toxic secondary products but also perturbs membrane architecture directly. Since the chemistry of membrane lipid containing PUFAs is prone to peroxidation reaction due to abstraction of H from methylene group (-CH2), it holds a solitary electron, leaving back an unpaired electron on the carbon ($-^{\bullet}CH_2$). Further, the presence of double bond in the fatty acid skeleton weakens the C-H bonds on the carbon atom near to the double bond and thus facilitates H subtraction. The preliminary reaction of most powerful ROS, $^{\bullet}OH$ with PUFA of membrane lipid, produces a lipid radical ($L^{\bullet}$) that in turn reacts with O_2, producing lipid peroxyl radical ($LOO^{\bullet}$), which subsequently can further abstract H from adjacent fatty acid to produce lipid hydroperoxides (LOOH) and a secondary lipid radical (Catala 2006). LOOH formed in this way may be cleared by transition metal ions, producing alkoxy radicals ($LO^{\bullet}$). Ultimately, generation of both peroxy and alkoxy radicals can stimulate propagation chain reaction of MLPO by abstracting other H atom (Buettner 1993). Therefore, the whole episode of MLPO perturbs the assembly of membrane components and their chemistry, causing changes in membrane fluidity and permeability which may be ultimately manifested in decontrolled molecular trafficking, loss of receptor function, ion homeostasis, and metabolic dysfunction (Nigam and Schewe 2000).

Enzymatic and nonenzymatic MLPO of PUFAs, one of the most vulnerable components to ROS, initiates oxidations that contribute in cascading autocatalytic chain reaction, causing extensive damage to membrane architecture. Due to MLPO several aldehydes are formed when lipid hydroperoxides are fragmented in cellular system. Out of the many, some of these secondary breakdown products are extremely reactive and, if allowed to accumulate, can induce toxicity. In fact, these secondary toxic chemicals can initiate further episodes of ROS-mediated oxidation events, when disseminated in cellular microclimate (Repetto et al. 2012). The secondary aldehydes produced by MLPO which are most extensively studied for their physiological relevance are HHE (4-hydroxy-2-hexenal), MDA (malondialdehyde), and HNE (4-hydroxy-2-nonenal). Out of these, HNE is recognized as the most significant aldehyde generated during MLPO of linoleic acid and arachidonic acid (n-6 PUFAs). HHE, on the other hand, is produced as a result of MLPO of linolenic acid and docosahexaenoic acid (n-3 PUFAs). Hence, 4-hydroxy-2-alkenals correspond to the most important aldehyde breakdown products generated during MLPO (Fig. 4.3). Physiological assessment of implication of accumulation of HNE and HHE revealed their cytotoxic function (Catala 2007).

HNE has been established as the most effective and reactive product of MLPO, having other physiological implications. Previous work shows that MLPO product HNE, apart from its cytotoxic effect under oxidative stress, exhibits intercellular redox signaling, modulating the expression of genes associated with cell proliferation and differentiation or event apoptosis (Repetto et al. 2010). The chemistry of HNE itself explains its high reactivity. HNE possesses hydroxyl groups close to its carbonyl groups, which enable them to react with thiol and amine groups of the target molecules they are encountering during oxidative stress (Repetto et al. 2010; Catala 2006). Its capability to diffuse in cellular environment further consolidates its signaling role. Further, the high reactivity of these secondary aldehydes causes interaction with major macromolecules of cell like DNA, protein, and phospholipid generating intra- and intermolecular adducts (Repetto et al. 2010). In fact, the oxidative protein damage associated with environmental stress is largely caused by HNE due to the addition of aldehyde compounds with –SH groups of cysteine, lysine amino groups, or histidine imidazole groups (Repetto et al. 2010).

Plant cell exhibits chemical devices to repair lipoperoxidized membrane by elimination of peroxidized fatty acid residues and their subsequent substitution by native fatty acids (Domingues et al. 2008; Lee et al. 1996). Several studies revealed that MLPO stimulates phospholipase A2 (PLA2)-mediated release of fatty acids necessary for their repair (Domingues et al. 2008; Lee et al. 1996).

Oxidative stress often triggers the formation of oxidized proteins, which in cellular system is being targeted hydrolyzed by 20S proteasomes (Jung et al. 2006). A very interesting finding, which strongly corroborates the coordinated function of proteasomal system and repair function of PLA_2 of peroxidized membrane, is the abundance of oxidized proteins near cell membrane. Therefore, simultaneous MLPO-mediated stimulation of PLA_2-mediated liberation of free fatty acids and activation of proteasomal system are conjugately responsible for replacement of peroxidized fatty acids with native fatty acids and proteins, causing membrane repair.

4.2.4 Some Secondary Products of MLPO Work as Reactive Lipid Species

MLPO are inherent features of cellular event that operate under normal environmental condition and aggravate significantly under unfavorable environmental cues and natural course of senescence. A major proportion of oxidized lipid peroxidation products, particularly formed by MLPO, happen to be electrophilic in nature and are extremely reactive and are competent enough to react with cellular nucleophiles like smaller peptides and amino acids like histidine, cysteine, lysine, etc. (Higden et al. 2012; Rudolph and Freeman 2009). With the growth of this domain of study related to MLPO, it is gradually evident that the process is not an expression of a chaotic event but seems to be endogenously regulated exhibiting multiple cellular functions, depending on the site of the event, condition, and mechanism of oxidation (Skorzynska-Polit 2007; Higden et al. 2012). Both enzymatic and nonenzymatic MLPO have many important biological implications (Niki et al. 2005). In fact, peroxidation of membrane-associated PUFAs results in the formation of RLS (reactive lipid species) which are electrophilic in nature (Fig. 4.4). As for example, peroxidation of arachidonic acid (a major substrate in MLPO process) results in the formation of several products, including a subset of electrophiles. Similarly, the nonenzymatic MLPO of PUFA also generate electrophiles like malondialdehyde (MDA), 4-hydroxynonenal (HNE), etc. (Davies et al. 2004).

Another source of RLS in plants is enzymatic MLPO, mediated by LOX. In plants, the commonly available LOXs are 9-LOX or 13-LOX, depending on the position of oxygenation of hydrocarbon backbone of free fatty acids of membrane phospholipid. Such oxygenation leads to the formation of 9(S) or 13(S) hydroperoxy derivatives of PUFAs, which ultimately get metabolized into secondary products that confer plant protection against pathogens (Gardner 1995; Gomi et al. 2002).

Several works proposed that LOX-mediated breakdown of an octadecadienoic acid, linoleic acid, is one of the probable mediators of defense mechanisms under biotic stress (Montillet et al. 2002). On the contrary, LOX-dependent MLPO

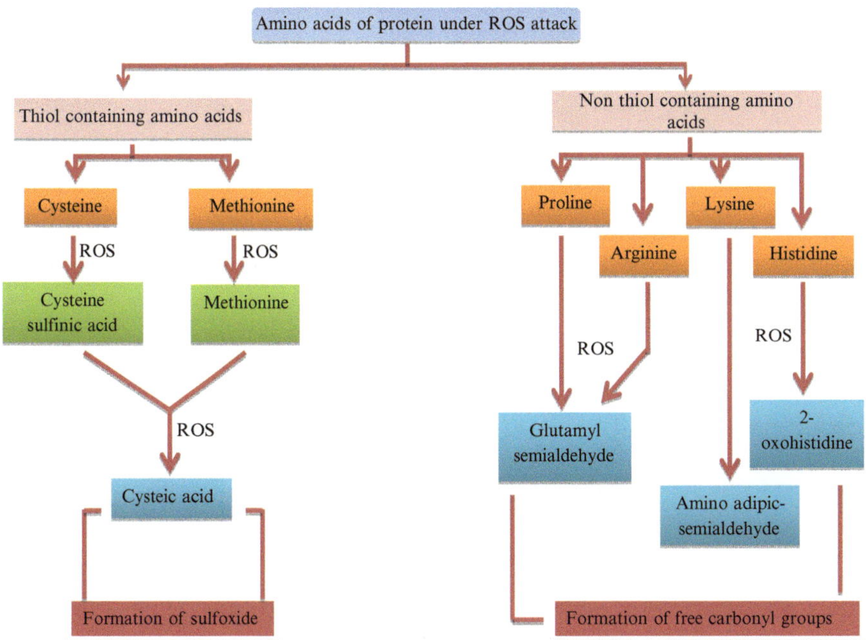

Fig. 4.4 Different types of oxidative modifications of amino acids

pathway has been shown to regulate hypersensitive reaction (HR) in plants as well (Montillet et al. 2002). While explaining these defense processes in plants through PCD or HR, it is found that a series of complex breakdown products formed during MLPO, some of which are RLS, exhibit a direct reactivity with some specific target molecules (Uchida 2003) and exert damaging effect. As discovered earlier, oxidation of PUFA generates MDA and hydroxyalkanals among other products which are RLS and form a variety of adducts with lysine residues of proteins or amine-containing heads of membrane phospholipids, causing damage (Uchida 2003; Kappus 1985).

In recent times, RLS have been found to take part in several important physiological events including cell death, induction of antioxidative defense, modification of cell signaling proteins, etc. (Higden et al. 2012; Fermer and Muller 2013). RLS can mediate biological responses basically either by irreversible covalent alteration of receptors or through reversible binding. In fact, in several animal systems, some RLS seems to work as ligands for specific receptors (Rudolph and Freeman 2009; Dickinson et al. 2006) and mediate biological effect through reversible receptor-ligand interaction initiating subsequent cell signaling event. Irreversible covalent modification of nucleophilic amino acid residues of proteins can also be mediated by RLS (Rudolph and Freeman 2009; Hong et al. 2005).

RLS-mediated episode of cell signaling can also take place through post-transitional modification of thiol-containing proteins, as evident during S-glutathionylation, disulfide formation, S-nitrosylation, etc. (Biswas et al. 2006; Ying et al. 2007). RLS usually targets proteins associated with cell signaling pathway and some other important redox-sensitive functional proteins like thioredoxin, thioredoxin reductase, HSP70, HSP90, 26S proteasome, k-Ras, cytochrome oxidase, etc. (Liu and Sok 2004; Biswas et al. 2006). Here the amino acid residue of target protein determines the specificity of reaction with RLS, which subsequently determine whether RLS evoke adaptive or cell death signaling response (Higden et al. 2012; Moran et al. 2006). One important example of RLS-induced adaptive response is "protein target/sensor" keap1, an adaptor protein associated with transcript factor Nrf2. In this case, the electrophilic attack of RLS causes translocation of transcription factor to nucleus, where it binds with ERF (electrophile response element) or ARF (antioxidative response element), upregulating stress acclamatory response.

4.3 Oxidative Modification to Protein

In order to maintain metabolic homeostasis, the functional conformations of proteins need to be preserved, and aggregation of conformationally changed proteins should be inhibited (Ghosh and Xu 2014; Anjum et al. 2014). Inhibition or dysfunction of protein metabolism is one of the prime and initial responses to any unfavorable environmental stress (Cramer et al. 2011; Timperio et al. 2008). However, there are ample evidences of the upregulation of synthesis of inducible proteins (HSPs, cold acclimation proteins, phytochelatins, LEA, dehydrins, MAP kinases, etc.), which help the plant maintain metabolic homeostasis and normal physiology under stress (Hossain et al. 2012; Fatehi et al. 2012; Ghosh and Xu 2014). However, owing to rapid rate of generation and high rate constants of a range of ROS, proteins are the target of oxidative deterioration among the cellular constituents under stress (Davies 2005; Anjum et al. 2014). As much of the three fourth of oxidized molecules in cell are found to be of protein (Rinalducci et al. 2008). The great abundance of oxidized protein in cell always makes the protein turnover rate very high. The oxidative modifications of the protein not only play important role in cell function but also in architectural abnormalities as well. The oxidative modification of several redox sensor proteins also have putative role in plant redox signaling. However, the oxidation of –SH components of the protein and the formation of carbonyl residues caused by environmental stresses often have served special role in ecotoxicological assessments.

4.3.1 Mechanism of Oxidative Modification of Cellular Proteins

ROS-induced oxidative modifications of constitutive aminoacids of protein fall into the following categories (Fig. 4.4):

(i) Modification of peptide bond
(ii) Direct oxidation of amino acids, mainly thiol-containing amino acids and formation of disulfide bonds
(iii) Oxidation of amino acid residues to form amino acid sulfoxide
(iv) Oxidation of amino acids and function of carbonyl groups in carbon skeleton (Dean et al. 1997; Rhoads et al. 2006)

Apart from these conventional protein oxidations, there are other processes of protein modification under oxidative stress, which include:

(i) Reaction of proteins with lipid peroxidation products
(ii) Protein reaction with RNS
(iii) ROS interaction with metallo-cofactors (Rhoads et al. 2006; Flint et al. 1993)

The common ROS which poses oxidative threat to protein includes $O_2^{\cdot-}$, H_2O_2, $\cdot OH$, $ROO\cdot$, $RO\cdot$, $HO_2\cdot$, $^1O_2$, etc. (Dean et al. 1997; Rhoads et al. 2006). Out of all these ROS, $\cdot OH$ is the most potent oxidant causing nonspecific oxidation of protein molecule (Davies 2005).

The direct modification of protein backbone under ROS attack through carbonylation, nitrosylation, disulfide linkage formation, adduct formation by glycoxidation, and glutathionylation influences the protein activity and subsequently the metabolism. ROS attack on protein however may be reversible as well (Ghezzi and Bonetto 2003). In fact, unlike irreversible oxidation, which causes permanent loss of protein function, reversible oxidation may have role in redox regulation. Carboxylation of lysine and arginine, nitration of tyrosine and tryptophan, and protein-protein cross-linking are irreversible in nature, whereas oxidative changes like nitrosylation and glutathionylation are reversible in nature.

Carbonylation of proteins and the loss of membrane protein thiol level (MPTL) are two important signs of irreversible oxidation processes that lead to loss of protein function and associated metabolic impairment. These two parameters are often considered as sensitive indicator of oxidative stress than MLPO, because the products of the latter may be catabolized more rapidly than oxidized proteins and hence are difficult to detect in right quantity (Palma et al. 2002). When compared, protein carbonylation (assessed in terms of accumulation of free carbonyl content) is found to be significantly more in the mitochondria as compared to chloroplasts and peroxisomes of leaves of cereal crops under stress, suggesting mitochondria as most susceptible organelle to oxidative protein damage (Bartoli et al. 2004). Some workers have identified a number of carbonylated soluble proteins from rice leaf mitochondria (Lam et al. 1989). Imbibitional extremes of temperature caused significant oxidative damage to protein in germinating tissues of two indica rice cultivars Ratna and SR26B, as evident from the accumulation of free carbonyl content and reduction in membrane protein thiol level, concomitant with elevated level of ROS level (Bhattacharjee 2013). Heavy metal treatment enhanced the level of carbonylation l from 4 to 5.6 nmol/mg protein in pea plants (Romero et al. 2002). The free carbonyl content also increases from 6.9 to 16.3 nmol/mg of peroxisomal protein under the exposure of heavy metal which might be due to the elevated ROS concentration in the peroxisomes (Romero et al. 2002). Several secondary lipid peroxidation products can also initiate oxidative damage to proteins. LPO products, such as 4-hydroxy-2-nonenal (HNE), can initiate oxidation of proteins under oxidative stress. Treatment of mitochondria either with ROS-generating chemical paraquat (upregulates superoxide formation in chloroplasts and mitochondria) or with secondary MLPO products like 4-hydroxy-2-nonenal (HNE) or even imposition of abiotic stresses leads to the oxidative modification of several mitochondrial enzymes (through covalent HNE-derived adduct of the lipoic acid moiety of target enzymes). The enzymes of TCA cycle and oxidative pentose phosphate pathway, like glycine decarboxylase, 2-oxoglutarate dehydrogenase, and pyruvate decarboxylase undergo oxidative modification under such conditions (Taylor et al. 2005; Millar and Lever 2000).

4.3.2 Abiotic Stress and Oxidative Modification of Proteins

In most of the cases, redox imbalance triggered by unfavorable environmental cues cause oxidation of proteins (Sharma et al. 2012; Miller et al. 2010). Heavy metals (Pb, Cd, Hg), salinity, and drought stresses cause protein oxidation (Stohs and Bagchi 1995; Tian et al. 2013; Gong et al. 2005). Heavy metals, in general, cause depletion of thiol-containing proteins and enhance carboxylated proteins (Roychoudhuri et al. 2012). Recently Bhoomika et al. (2014) reported different levels of carboxylated proteins in roots and shoots of rice showing different Al tolerance. Comparison of heavy metal-tolerant and susceptible cultivars of maize under heavy metal exposure also shows differential accumulation of RC=O. The differential susceptibility of protein in these germplasms is basically due to different levels of oxidative stress under heavy metal exposure (Stadtman 1992).

Chakraborty and Bhattacharjee (2015) reported significantly higher accumulation of RC=O in imbibitional chilling and oxidative stress-raised seedlings of salinity-resistant rice cultivar (SR26B) as compared to salinity-susceptible cultivar (Ratna). The lesser extent of protein oxidation of salinity-resistant cultivar under extremes of temperature may be one of the reasons of cross-tolerance of rice cultivar (Chakraborty and Bhattacharjee 2015). Similarly, Tian et al. (2013) reported drought stress-mediated enhancement in protein carboxylation in leaves of mutant (*tasg1*) and wild cultivars of wheat. The greater functional stability of thylakoid membrane proteins tasg1 as compared to wild type found to play important role in drought tolerance. The differential level of RC=O between rice germplasms differing in sensitivity toward NaCl salinity (Gobindobhog and Nona Bokra) may be the diagnostic feature of salt tolerance in rice (Roychoudhuri et al. 2011).

4.4 Cross Talk Between ROS-Mediated Protein Oxidation and Lipid Peroxidation

Peroxidation of membrane lipids always causes formation of a great number of secondary products including malondialdehyde, oxylipins, and smaller lipid-derived reactive electrophiles (Almaras et al. 2003). There are evidences of the MLPO products reacting with proteins, other lipids, and nucleic acids (Rhoads et al. 2006). Further, MDA and HNE can form additive complexes and initiate protein damage.

In fact, conjugation of aldehyde with protein can bring indirect oxidation of protein through carboxylation (Moller et al. 2011). Moreover, oxidative deterioration of PUFA initiates chain reaction, causing formation of cytotoxic carbonyl compounds like α,β-unsaturated aldehydes, dialdehydes, and ketoaldehydes which may subsequently get attached with proteins due to their higher reactivity. In a model experiment, Rubisco or BSA as model protein was found to react with C18PUFA (the MDA generated form of peroxidized linolenic acid) leading to the modifications of proteins (Yamauchi et al. 2008). Similarly, the LHCPS of photosynthetic apparatus was found to be modified by MDA under EPE as heat stress in *Arabidopsis* (Millar and Leaver 2000). The MLPO products like 4-hydroxynonenal are found to be extremely active for targeting proteins and subsequently causing damage or loss of function (Taylor et al. 2005).

4.5 Assessment of Products of MLPO and PO as Sensitive Redox Biomarkers for the Evaluation of Impact of Environmental Stress

Membrane lipid peroxidation, an event primarily upregulated under environmental stress-induced oxidative stress, can be assessed by monitoring changes in the level of peroxidation products or accumulation of secondary oxidative products (Frankel 1991; Halliwell and Whiteman 2004). The assessment of loss of membrane phospholipid-bound unsaturated fatty acids can also be one of the sensitive parameters for estimation of MLPO (Halliwell and Chirico 1993). However, out of several associated processes, monitoring the titer of the end products of MLPO is the most widely accepted procedure for the estimation of oxidative stress and associated oxidative damage of the plant tissue (Shulaev and Oiver 2006). Environmental stress-associated loss of redox homeostasis due to overaccumulation of ROS causes peroxidation of PUFAs, producing α,β-unsaturated aldehydes like 4-hydroxyenal, malondialdehyde, etc. (Hartley et al. 1999). These products of MLPO are the accepted markers of oxidative stress of plants (Frankel 1991). Several analytical and biochemical techniques can be exploited to assay MLPO for understanding the changes in redox status or magnitude of oxidative stress suffered by plants under stress (Table 4.1).

Estimation of thiobarbituric acid reactive substances (TBARS) for the assessment of accumulation of malondialdehyde, a product of MLPO, is based on its reaction with TBA followed by measuring absorbance at 532 nm (Draper and Hadley 1990; Hodges et al. 1999). Assay TBARS and its modification are used as a sensitive test of MLPO for assessing the impact of abiotic stress (temperature, drought, heavy metal, salinity, UV irradiation stress) of plants (Bhattacharjee 2008; Bhattacharjee and Mukherjee 2002, 2004; Mahan and Mauget 2005). Although used widely for assessing MLPO and oxidative stress of plants, there are some drawbacks of this process, where there may be formation of thiobarbituric acid reactive substances (TBARS) which are not related to MLPO (Halliwell and Whiteman 2004). In fact, TBA test works well with different membrane system like microsomal and liposomal membrane system, but its application with other membrane preparation poses problem (Gutteridge 1986). Aldehydes other than MDA can also take part in the reaction and form chromogens with same absorbance at 532 nm as TBARS test (Kosugi et al. 1987). A serious problem associated with TBARS test is the instability of secondary aldehydes as they fast got converted to other products. Another weakness of the process is that TBA test rarely measures free MDA content of the system; instead most if not all MDA measured is generated by decomposition of lipid peroxides during acid heating of the test (Esterbauer et al. 1991). On top of these, the peroxide decomposition produces radicals that can start peroxidation of other fatty acids during the assay, thereby causing amplified response (Gutteridge and Tickner 1978). Endogenous H_2O_2 were also found to exert false-positive results when TBA test is applied to microsomal membrane (Cecchini et al. 1990).

Recent progress of MS techniques caused the development of more accurate GC-MS bound-based process for the estimation of 4-HNE and MDA (Liu et al. 1997; Muckenschnabel et al. 2001). A highly sensitive LC-MS-based detection of 2,4-dinitrophenyl hydrazine derivative of 4HNE and MDA has been practiced since the late 1990s (Deighton et al. 1997; Muckenschnabel et al. 2002). One significant advantage of LC-MS and GC-MS bound test is the ability to identify individual lipid species targeted by ROS (Byrdwell and Neff 2002).

The structures of conjugated diene, a product of MLPO, absorb UV in the range of 230–235 nm. Therefore, assessment of absorbance at 230–235 nm of tissue leachate containing the products of lipid peroxidation can enable us to indirectly

Table 4.1 Common methods of detection and estimation of membrane lipid peroxidation

Procedure	Test and the sensitivity of the process	References
1. Thiobarbituric acid test	Thiobarbituric acid reactive substances (TBARS) like different aldehydes, malondialdehyde, react with TBA at low p^H and form [TBA]-MDA adduct (pink chromogen with absorbance at 532 nm). Standardized test, sensitive for microsomal, liposomal MLPO test Rarely measures the free MDA content of lipid system. TBA reactivity depends on the lipid content of the sample	Halliwell and Whiteman (2004) and Gutteridge (1988)
2. Fluorescence test for aldehyde and end product of MLPO	MDA and other aldehydes, the end products of MLPO, are measured. Aldehydes react with -NH2 group forming Schiff s bases in acidic p^H. At neutral p^H fluorescence dihydropyridines are formed. Aldehydes got polymerized to form fluorescence products and fluorescence is measured. Highly sensitive method, although formation of fluorescence products takes place in a minor complex reaction	Halliwell and Chirico (1993) and Porta (1991)
3. Diene conjugation test	Diene-conjugated structures, formed by oxidation of PUFAs as the intermediates of MLPO, are assessed. Absorbance in the UV range (230–235 nm) is useful for the measurement of MLPO of pure lipids. The process requires specific extraction techniques	Halliwell and Chirico (1993) and Corongiu et al. (1986)
4. LC-MS-/GC-MS-based techniques	Aldehydes and lipid peroxides produced by MLPO are identified and estimated. DNPH derivatives of MDA/4HNE have been utilized in these processes. Peroxidation products are extracted and reduced to alcohols, which are subsequently separated and identified by MS techniques	Finkel et al. (1989) and Liu et al. (1997)
5. HPLC-based process	Estimation of loss of fatty acids by MLPO. Extremely useful for the estimation of LPO stimulated by metal ions	Frei et al. (1991) and Halliwell and Whiteman (2004)
6. Spin-trapping method	Intermediate radical species formed by MLPO are identified	Janzen (1990)
7. Glutathione peroxidize (GPX) method	GPX made to react with hydroperoxides and H_2O_2 forming oxidized glutathione (GSSG) which subsequently react with NADPH to regenerate reduced glutathione (GSH). Consumption of NADPH in the reaction is estimated as the rate of LPO. However, membrane-associated peroxide needs to be separated first for the estimation of MLPO	Halliwell and Chirico (1993)

(continued)

Table 4.1 (continued)

Procedure	Test and the sensitivity of the process	References
8. UV-spectrophotometric test for free carbonyl content	Oxidative damage to proteins was estimated as the content of carbonyl groups. Degraded proteins were made to react with potassium phosphate buffer (pH 7.0) containing EDTA, PMSF, DTT and leupeptin, aprotinin, and antipain (protease inhibitors) and centrifuged and finally made to react with DNPH. It is followed by TCA precipitation and suspension of precipitated protein in guanidine hydrochloride and estimation of RC=O by reading absorbance at 370 nm	Jiang and Zhang (2001)
9. DNPH-associated assays for carbonyls	Utilize anti-DNP antibodies to quantify carbonyl content	Buss et al. (1997)
10. Other immunochemical techniques	Total carbonyls can be measured by ELISA, slot blotting, immunohistochemistry	Smith et al. (1998)

measure MLPO (Jack et al. 1991). Several workers used HPLC to separate further the UV-absorbing diene conjugates for further analysis (Dormandy and Wickens 1987) and estimation of membrane damage. Based on existing assay techniques for the assessment of extent and magnitude of MLPO that a tissue suffers, it is significant to employ techniques that not only extend precise chemical information about the products of the process but also can adeptly quantify the products of the said process, regardless of whether short or long lived. Thus, more emphasis has to be given on separation technique and sensitivity of MLPO products. These are often achieved by HPLC (assaying peroxides, aldehydes)-, GC (assaying hexanals)-, and LC (assaying dienes)-based techniques (Liu et al. 1997; Frankel et al. 1989; Thomas et al. 1991). However, for these to achieve, the primary goal should be the efficient sample preparation with extreme care so as to ensure minimum loss of oxidized material and at the same time artificial peroxidation can be avoided during the process.

4.6 Oxidative Damage to Nucleic Acids

Though in general the plant genome is very stable, its DNA might get damaged due to the exposure to environmental stress which might damage the DNA by instigating a secondary oxidative stress and thereby exerts genotoxic effect. The most potent ROS which induce the endogenous damage to DNA, commonly known as "spontaneous DNA damage" are $OH^{\cdot}$ and $O_2^{\cdot-}$. Loss of redox homeostasis, which causes accumulation of high levels of ROS, can cause damage to nucleic acids. It has been reported that $OH^{\cdot}$ is the most reactive one and cause damage to all components of the DNA molecule, damaging both the purine and pyrimidine bases and also the deoxyribose backbone, $^1O_2$ primarily attacks guanine, and H_2O_2 and $O_2^{\cdot-}$ don't react at all. ROS thus is capable of inducing damage to DNA which includes:

(i) Deletion of nucleic acid base
(ii) Cross-linking, formation of pyrimidine dimers, and strand breaks
(iii) Base modification, such as oxidation and alkylation

Oxidative damage to DNA results in various molecular and physiological effects, such as downregulation of protein synthesis, membrane damage, and damage to LHCPs and photosynthetic redox proteins, which affect growth and development of the whole organism. DNA damage can result either in arrest or induction of transcription, change in signal transduction pathways, replication errors, and genomic instability. The DNA damage caused by exposure to UV-B is mainly due to the formation of dimers between adjacent pyrimidines, forming UV photoproducts like cyclobutane pyrimidine dimmers (CPDs) and 6-4PPs dimers. Both MLPO and oxidative DNA damage have been considered major determinants of loss of seed viability and also associated with senescence.

4.7 Conclusion and Perspective

The loss of redox homeostasis, particularly under environmental stresses, instigates oxidative deterioration of important biomolecules, primarily mediated by prooxidants. The most important cellular damages are marked as lipid peroxidation, protein oxidation, and nucleic acid damage. Both PO and LPO are natural and essential process and mechanistically important as both processes not only involve oxidative deterioration but may also have significant role in cellular physiology. Decontrolled oxidative damages to important cellular molecules are associated with senescence, aging, and environmental stress and exert cytotoxic effect on cells, whereas the regulated processes, particularly MLPO and PO, function as effectors. However, evolution has nominated or co-opted the MLPO to provide some important cellular functions like intercellular communication, understanding environmental cues, programmed cell death, modification of biophysical structure of membrane, etc. Though the process of MLPO seems to be apparently deleterious and associated with the formation of a potential source for ROS, a serious look into the process reveals several physiological significance and relevance. In one hand it is associated with repair of damaged membrane and capable of initiating HR; on the other hand, it can modify membrane-associated lipid-lipid interaction capable of changing fluidity of membranes. Nevertheless, the most noteworthy role of the process is the contribution of RLS, capable of modulation of cell signaling and physiology. The development of sensitive mass spectrometric technique is required for the identification and characterization of oxylipidomes, an important subset of lipidomes formed through MLPO under oxidative stress. Development of techniques to monitor specific RLS-protein (sensor) adducts to define electrophile-responding proteome for specific RLS will help us to understand the paradigm of cell signaling mechanisms. It is now clear that RLS react with discreet electrophile-responsive proteome and subsequently sponsor signaling events which evokes adaptive response.

In summary, the current paradigm is that how the cells exploit MLPO and their products particularly under different physiological conditions. Accumulation of different levels of RLS plays crucial role as low titer of RLS can modify specific target protein (sensor) in contrary to high titer where RLS can modify other cellular macromolecules in less specific manner. Thus, failure to control or fine-tune the process always leads to deteriorative events and ultimately to lethal consequence. So, it is extremely important to critically unfold the process of regulation of MLPO and identification of specific cellular targets of RLS in the physiology of organisms. It would be extremely interesting and challenging to identify changes in gene expression initiated by MLPO through RLS. Such a global analysis of effect of RLS on transcriptome of plants has not yet been attained, but with emergence of post-genomic technologies, it will be not far off. In order to gain further understanding of stress-induced changes in oxidative deterioration and their physiological consequences in plants, further research should be planned to identify and characterize it.

References

Alméras E, Stolz S, Vollenweider S, Reymond P, Mène-Saffrane L, Farmer EE (2003) Reactive electrophile species activate defense gene expression in *Arabidopsis*. Plant J 34:205–216

Anjum NA, Gill SS, Gill R, Hasanuzzaman M, Duarte AC, Pereira E (2014) Metal/metalloid stress tolerance in plants: role of ascorbate, its redox couple, and associated enzymes. Protoplasma 251:1265–1283. https://doi.org/10.1007/s00709-014-0636-x

Aust SD, Moorehouseand CE, Thomas CE (1995) Role of metals in oxygen radical reactions. J Free Radic Biol Med 01:03–25

Bartoli CG, Gomez F, Martinez DE, Guiamet JJ (2004) Mitochondria are the main target for oxidative damage in leaves of wheat (*Triticum aestivum* L.). J Exp Bot 55:1663–1669

Bhattacharjee S (2005) Reactive oxygen species and oxidative burst: roles in stress, senescence and signal transduction in plants. Curr Sci 89:1113–1121

Bhattacharjee S (2008) Calcium –dependent signaling pathway in the heat induced oxidative injury in *Amaranthus lividus* L. Biol Plant 52(01):137–140

Bhattacharjee S (2012a) An inductive pulse of hydrogen peroxide pretreatment restores redox-homeostasis and mitigates oxidative membrane damage under extremes of temperature in two rice cultivars (*Oryza sativa* L., Cultivars Ratna and SR 26B). Plant Growth Regul 68:395–410

Bhattacharjee S (2012b) The language of reactive oxygen species signaling in plants. J Bot 2012:22. https://doi.org/10.1155/2012/985282.01-22

Bhattacharjee S (2013) Heat and chilling induced disruption of redox homeostasis and its regulation by hydrogen peroxide in rice (*Oryza sativa* L., Cultivar Ratna). Physiol Mol Biol Plants 19:199–207

Bhattacharjee S (2014) Membrane lipid peroxidation and its conflict of interest: the two faces of oxidative stress. Curr Sci 107:1811–1823

Bhattacharjee S, Mukherjee AK (2002) Salt stress induced cytosolute accumulation, antioxidant response and membrane deterioration in three rice cultivars during germination. Seed Sci Technol ISTA 30:279–287

Bhattacharjee S, Mukherjee AK (2004) Heavy metal induced germination and early growth impairment in *Amaranthus lividus* L.: implications of oxidative membrane damage. J Plant Biol 31(1):01–11

Bhoomika K, Pyngrope S, Dubey RS (2014) Effect of aluminum on protein oxidation, non-protein thiols and protease activity in seedlings of rice cultivars differing in aluminum tolerance. J Plant Physiol 171:497–508

Biswas S, Chida AS, Rahman I (2006) Redox modifications of protein-thiols: emerging roles in cell signaling. Biochem Pharmacol 71:551–564

Boveris A, Repetto MG, Bustamante J, Boveris AD, Valdez LB (2008) The concept of oxidative stress in pathology. In: Álvarez S, Evelson P (eds) Free radical pathophysiology. Transworld Research Network, Kerala, pp 1–17. ISBN: 978-81-7895-311-3

Buettner GR (1993) The pecking order of free radicals and antioxidants: lipid peroxidation, alpha-tocopherol, and ascorbate. Arch Biochem Biophys 300:535–543

Buss H, Chau PT, Scuis KB, Domigan NM, Winterbourn CC (1997) Protein carbonyl measurement by a sensitive ELISA method. Free Radic Biol Med 23(3):361–363

Byrdwell WC, Neff WE (2002) Dual parallel electrospray ionization and atmospheric pressure chemical ionization mass spectrometry (MS), MS/MS and MS/MS/MS for the analysis of triacylglycerols and triacylglycerol oxidation products. Rapid Commun Mass Spectrom 16:300–319

Catala A (2006) An overview of lipid peroxidation with emphasis on outer segments of photoreceptors and the chemiluminescence assay. Int J Biochem Cell Biol 38:1482–1495

Catala A (2007) The ability of melatonin to counteract lipid peroxidation in biological membranes. Curr Mol Med 7:638–649

Cecchini R, Aruoma OI, Halliwell B (1990) The action of hydrogen peroxide on the formation of thiobarbituric acid reactive material from microsomes, liposomes or from DNA damaged by bleomycin or phenanthroline: artefacts in the thiobarbituric acid test. Free Radic Res Commun 10:245–258

Chakraborty A, Bhattacharjee S (2015) Differential competence of redox-regulatory mechanism under extremes of temperature determines growth performances and cross tolerance in two indica rice cultivars. J Plant Physiol 176:65–77

Codreanu SG, Zhang B, Sobecki SM, Billheimer DD, Liebler DC (2009) Global analysis of protein damage by the lipid electrophile 4-hydroxy-2-nonenal. Mol Cell Proteomics 8:670–680

Corongiu FP, Poli G, Dianzani MU (1986) Lipid peroxidation and molecular damage to polyunsaturated fatty acids in rat liver: recognition of two classes of hydroperoxides formed under conditions in vivo. Chem Biol Interact 59:147–155

Cramer GR, Urano K, Delrot S, Pezzotti M, Shinozaki K (2011) Effects of abiotic stress on plants: a systems biology perspective. BMC Plant Biol 11:163

Davies MJ (2005) The oxidative environment and protein damage. Biochim Biophys Acta 1703:93–109

Davies SS, Amarnath V, Roberts LJ (2004) Isoketals: highly reactive γ-ketoaldehydes formed from the H2–isoprostane pathway. Chem Phys Lipids 128:85–99

Dean RT, Fu S, Stocker R, Davies MJ (1997) Biochemistry and pathology of radical-mediated protein oxidation. Biochem J 324:1–18

deGroot JJMC, Veldink GA, Velegenthart JFC, Boldingh J, Weverand R, vanGelder BF (1975) Demonstration of EPR spectroscopy of the functional role of iron in soybean lipoxygenase. J Biochem Biophys Acta 377:71–79

Deighton N, Magill WJ, Bremner DH, Benson EE (1997) Malondialdehyde and 4-hydroxy-2-nonenal in plant tissue cultures: LC-MS determination of 2,4-dinitrophenylhydrazone derivatives. Free Radic Res 27:255–265

Dianzani M, Barrera G (2008) Pathology and physiology of lipid peroxidation and its carbonyl products. In: Álvarez S, Evelson P (eds) Free radical pathophysiology. Transworld Research Network, Kerala, pp 19–38. ISBN: 978-81-7895-311-3

Dickinson DA, Darley-Usmar VM, Landar A (2006) The covalent advantage: a new paradigm for cell signaling by thiol reactive lipid oxidation products. In: Dalle-Donne I, Scalone A, Butterfield DA (eds) Redox proteomics: from protein modifications to cellular dysfunction and diseases. Wiley, Indianapolis, pp 345–367

Domingues MR, Reis A, Domingues P (2008) Mass spectrometry analysis of oxidized phospholipids. Chem Phys Lipids 156:1–12

Dormandy TL, Wickens DG (1987) The experimental and clinical pathology of diene conjugation. Chem Phys Lipids 45:356–364

Draper HH, Hadley M (1990) Malondialdehyde determination as index of lipid peroxidation. Methods Enzymol 186:421–431

Esterbauer H, Schaur RG, Zollner H (1991) Chemistry and biochemistry of 4-hydroxynonenal, malondialdehyde and related aldehydes. Free Radic Biol Med 11:81–128

Farooqui T, Farooqui A (2011) Lipid-mediated oxidative stress and inflammation in the pathogenesis of Parkinson's disease. Park Dis 2011:247467. https://doi.org/10.4061/2011/247467

Fatehi F, Hosseinzadeh A, Alizadeh H, Brimavandi T, Struik P (2012) The proteome response of salt-resistant and salt-sensitive barley genotypes to long-term salinity stress. Mol Biol Rep 39:6387–6397

Fermer EF, Muller MJ (2013) Reactive oxygen species mediated lipid peroxidation and RES activated signaling. Annu Rev Plant Biol 04:429–450

Feussner I, Wasternack C (2004) The lipoxygenase pathway. Annu Rev Plant Biol 53:275–297

Feussner I, Balkenhohl TJ, Porzel A, Kuhn H, Wasternack C (1997) Structural elucidation of oxygenated storage lipids in cucumber cotyledons- implication of lipid body lipoxygenase in lipid mobilization during germination. J Biol Chem 272:21635–21641

Flint DH, Tuminello JF, Emptage MH (1993) The inactivation of Fe–S cluster containing hydrolyases by superoxide. J Biol Chem 268:22369–22376

Foyer CH, Noctor G (2013) Redox signaling in plants. Antioxid Redox Signal 18(16):2087–2096

Foyer CH, Shigeoka S (2011) Understanding oxidative stress and antioxidant functions to enhance photosynthesis. Plant Physiol 155:93–100

Frankel EL (1991) Recent advances in lipid oxidation. J Sci Food Agric 54:495–511

Frankel EN, Hu ML, Tappel AL (1989) Rapid headspace gas chromatography of hexanal as a measure of lipid peroxidation in biological samples. Lipids 24:976–981

Frei B, Fort TM, Ames B, Cross CE (1991) Gas phase oxidants of cigarette smoke induce lipid peroxidation and changes in lipoprotein properties in human blood plasma. Biochem J 277:133–138

Fritz KS, Petersen DR (2011) Exploring the biology of lipid peroxidation-derived protein carbonylation. Chem Res Toxicol 24:1411–1419

Fuchs C, Spiteller G (2000) Iron release from the active site of the lipoxygenase. Z Naturforsch 55C:643–649

Fuller AM, Weichert H, Fischer M, Feussner I, Grimes DH (2001) Activity of soybean lipoxygenase isoforms against esterified fatty acid indicates functional specificity. Arch Biochem Biophys 6:146–154

Gardner HW (1995) Biological roles and biochemistry of the lipoxygenase pathway. Hortscience 30:197–205

Ghezzi P, Bonetto V (2003) Redox proteomics: identification of oxidatively modified proteins. Proteomics 3:1145–1153

Ghosh D, Xu J (2014) Abiotic stress responses in plant roots: a proteomics perspective. Front Plant Sci 5:6. https://doi.org/10.3389/fpls.2014.00006

Gill SS, Tuteja N (2010) Reactive oxygen species and antioxidant machinery in abiotic stress tolerance in crop plants. Plant Physiol Biochem 48:909–930

Gomi K, Yamamoto H, Akimitsu K (2002) Characterization of a lipoxygenase gene in rough lemon induced by *Alternaria alternata*. Gen Plant Pathol 68:21–30

Gong R, Dingh Liu Y, Chen Q, Liu H (2005) Lead biosorption and desorption by intact and pretreated *Spirulina maxima* biomass. Chemosphere 58:125–130

Gutteridge JMC (1986) Aspects to consider when detecting and measuring lipid peroxidation. Free Radic Res Commun 1:173–184

Gutteridge JMC (1988) Lipid peroxidation: some problems and concepts. In: Halliwell B (ed) Oxygen radicals and tissue injury. Allen Press, Lawrence, pp 9–19. 71

Gutteridge JMC, Tickner TR (1978) The thiobarbituric acid- reactivity of bile pigments. Biochem Med 19:127–132

Halliwell B, Chirico S (1993) Lipid peroxidation: its mechanism measurement and significance. Am J Clin Nutr 57:715s–725S

Halliwell B, Gutteridge JMC (1999) Free radical in biology and medicine, 3rd edn. Oxford University Press, Oxford

Halliwell B, Whiteman M (2004) Measuring reactive species and oxidative damage in vivo and in cell culture: how should you do it and what do the results mean? Br J Pharmacol 142:231–255

Hartley DP, Kolaja KL, Richard J, Petersen DR (1999) 4-Hydroxynonenal and malondialdehyde hepatic protein adducts in rats treated with carbon tetrachloride immunochemical detection and lobular localization. Toxicol Appl Pharmacol 161:23–33

Higden A, Dlvers AR, Jyoh J, Lender A, Darlayusmar VM (2012) Cell signaling by reactive lipid species: new concept and molecular mechanism. Biochem J 442:453–464

Hodges DM, DeLong JM, Forney CF, Prange RK (1999) Improving the thiobarbituric acid-reactive-substances assay for estimating lipid peroxidation in plant tissues containing anthocyanin and other interfering compounds. Planta 207:604–611

Hong F, Sekhar KR, Freeman ML, Liebler DC (2005) Specific patterns of electrophile adduction trigger Keap 1 ubiquitination and Nrf2 activation. J Biol Chem 280:31768–31775

Hossain Z, Nouri M-Z, Komatsu S (2012) Plant cell organelle proteomics in response to abiotic stress. J Proteome Res 11:37–48

Jabs T (1999) Reactive oxygen intermediates as mediators of programmed cell death in plants and animals. Biochem Pharmacol 57:231–245

Jack CIA, Jackson MJ, Ridgeway E, Hind CRK (1991) Octadeca-9,11-dienoic acid-a measurement of free radical activity or a marker of infection in the lung? Clin Sci 81:17–21

Janzen EG (1990) Spin trapping and associated vocabulary. Free Radic Res Commun 10:63–68

Jiang M, Zhang J (2001) Effect of abscisic acid on active oxygen species, antioxidative defense system and oxidative damage in leaves of maize seedlings. Plant Cell Physiol 42(11):1265–1273

Jung T, Engles M, Kaiser B, Poppek D, Grune T (2006) Intracellular distribution of oxidized proteins and proteasome in HT22 cells during oxidative stress. Free Radic Biol Med 40:1303–1312

Kappus H (1985) Lipid peroxidation: mechanism, analysis, enzymology and biological relevance. In: Sies H (ed) Oxidative stress. Academic, London, pp 273–310

Kosugi H, Kato T, Kikugawa K (1987) Formation of yellow, orange and red pigments in the reaction of alk-2-enals with 2-thiobarbituric acid. Anal Biochem 165:456–464

Krasensky J, Jonak C (2012) Drought, salt, and temperature stress-induced metabolic rearrangements and regulatory networks. J Exp Bot 63:1593–1608

Lam E, Benfey PN, Gilmartin PM, Fang RX, Chua NH (1989) Site-specific mutations alter in vitro factor binding and change promoter expression pattern in transgenic plants. Proc Natl Acad Sci U S A 86:7890–8104

Landar A, Giles NM, Zmijewski JW, Watanabe N, Oh JY, Darley-Usmar VM (2006) Modification of lipids by reactive oxygen and nitrogen species: the oxy-nitroxy-lipidome and its role in redox cell signaling. Futur Lipidol 1:203–211

Lee YC, Zheng YO, Taraschi TF, Janes N (1996) Hydrophobic alkyl head groups strongly promote membrane curvature and violate the head group volume correlation due to "head group" insertion. Biochemistry 35:3677–3684

Liu XW, Sok DE (2004) Inactivation of protein disulfide isomerase by alkylators including α,β-unsaturated aldehydes at low physiological pHs. Biol Chem 385:633–637

Liu J, Yeo HC, Doniger SJ, Ames BN (1997) Assay of aldehydes from lipid peroxidation: gas chromatography-mass spectrometry compared to thiobarbituric acid. Anal Biochem 245:161–166

Mahan JR, Mauget SA (2005) Antioxidant metabolism in cotton seedlings exposed to temperature stress in the field. Crop Sci 45:2337–2345

Marwah SS, Blann AD, Rea C, Phillips JD, Wright J, Bareford D (2002) Reduced vitamin E antioxidant capacity in sickle cell disease is related to transfusion status but not to sickle crisis. Am J Hematol 69:144–146

Millar AH, Leaver CJ (2000) The cytotoxic lipid peroxidation product, 4-hydroxy-2-nonenal, specifically inhibits decarboxylating dehydrogenases in the matrix of plant mitochondria. FEBS Lett 481:117–121

Miller G, Suzuki N, Ciftci-Yilmaz S, Mittler R (2010) Reactive oxygen species homeostasis and signalling during drought and salinity stresses. Plant Cell Environ 33:453–467. https://doi.org/10.1111/j.1365-3040.2009.02041.x
Mittler R (2002) Oxidative stress, antioxidants and stress tolerance. Trends Plant Sci 7:405–410
Møller IM, Rogowska-Wrzesinska A, Rao RS (2011) Protein carbonylation and metal-catalyzed protein oxidation in a cellular perspective. J Proteome 74:2228–2242
Montillet J-L, Agnel J-P, Ponchet M, Vailleau F, Roby D, Trantaphylides C (2002) Lipoxygenase-mediated production of fatty acid peroxides is a specific signature of the hypersensitive reaction in plants. Plant Physiol Biochem 40:633–639
Moran LK, Gutteridge JM, Rahman I (2006) Redox modification of protein thiols: emerging roles in cell signaling. Curr Med Chem 08:763–772
Morrow JD (2005) Quantification of isoprostanes as indices of oxidant stress and the risk of atherosclerosis in humans. Arterioscler Thromb Vasc Biol 25:279–286
Muckenschnabel I, Williamson B, Goodman BA, Lyon GD, Stewart D, Deighton N (2001) Markers for oxidative stress associated with soft rots in French beans (*Phaseolus vulgaris*) infected by *Botrytis cinerea*. Planta 212:376–381
Muckenschnabel I, Goodman BA, Williamson B, Lyon GD, Deighton N (2002) Infection of leaves of *Arabidopsis thaliana* by *Botrytis cinerea*: changes in ascorbic acid, free radicals and lipid peroxidation products. J Exp Bot 53:207–214
Nigam S, Schewe T (2000) Phospholipase A2s and lipid peroxidation. Biochem Biophys Acta 1488:167–181
Niki E, Yoshida Y, Saito Y, Noguchi N (2005) Lipid peroxidation: mechanisms, inhibition, and biological effects. Biochem Biophys Res Commun 338:668–676
Palma JM, Sandalio LM, Corpas FJ, Romero-Puertas MC, Mccarthy I, Del Rio LA (2002) Plant proteases, protein degradation, and oxidative stress: role of peroxisomes. Plant Physiol Biochem 40(6–8):521–530
Porta EA (1991) Advantages in age pigment research. Arch Gerontol Geriatr 12:303–320
Porta H, Rocha-Sosa M (2002) Plant lipoxygenases, physiological and molecular features. Plant Physiol 130:15–21
Pralico D, Lawsen JA, Rostach J, Fitzgovald GA (2001) The isoprostane. Biol Med Trends Endocrinol Metab 12:243–247
Repetto M (2008) Clinical use of chemiluminiscence assays for the determination of systemic oxidative stress. In: Propov I, Lewin G (eds) Handbook of chemiluminescence methods in oxidative stress assessment. Transworld Research Network, Kerala, pp 433–450
Repetto MG, Boveris A (2012) Transition metals: bioinorganic and redox reactions in biological systems. In: Nova Science Publishers Inc (ed) Transition metals: uses and characteristics. Nova Science Publishers Inc, New York, pp 349–370. ISBN: 978-1-61761-110-0
Repetto MG, Ferrarotti NF, Boveris A (2010) The involvement of transition metal ions on iron-dependent lipid peroxidation. Arch Toxicol 84:255–262. ISSN: 0340-5761
Repetto QM, Seprina J, Boveris A (2012) Lipid peroxidation: chemical mechanism, biological implication and analytical determination, Intech, pp 03–28. https://doi.org/10.5772/45923
Rhoads DM, Umbach AL, Subbaiah CC, Siedow JN (2006) Mitochondrial reactive oxygen species—contribution to oxidative stress and interorganellar signaling. Plant Physiol 141:357–366
Rinalducci S, Murgiano L, Zolla L (2008) Redox proteomics: basic principles and future perspectives for the detection of protein oxidation in plants. J Exp Bot 59:3781–3801
Romero Puertas M, Palma J, Gómez M, Del Rio L, Sandalio L (2002) Cadmium causes the oxidative modification of proteins in pea plants. Plant Cell Environ 25:677–686
Roychoudhury A, Basu S, Sengupta DN (2011) Amelioration of salinity stress by exogenously applied spermidine or spermine in three varieties of indica rice differing in their level of salt tolerance. J Plant Physiol 168:317–328
Roychoudhury A, Basu S, Sengupta DN (2012) Antioxidants and stress-related metabolites in the seedlings of two indica rice varieties exposed to cadmium chloride toxicity. Acta Physiol Plant 34(3):835–847. https://doi.org/10.1007/s11738-011-0881-y

Rudolph TK, Freeman BA (2009) Transduction of redox signaling by electrophile-protein reactions. Sci Signal 2:7–11

Sharma P, Jha AB, Dubey RS, Pessarakli M (2012) Reactive oxygen species, oxidative damage, and antioxidative defense mechanism in plants under stressful conditions. J Bot 2012:217037

Shulaev V, Oiver DJ (2006) Metabolic and proteomic marker for oxidative stress. New tools of reactive oxygen species research. Plant Physiol 141:367–372

Skorzynska-Polit E (2007) Lipid peroxidation in plant cell, its physiological role and changes under heavy metal stress. Acta Soc Bot Pol 76(01):49–54

Smith MA, Sayre LM, Anderson VE, Harris PLR, Beal MF, Kowall N, Perry G (1998) Cytochemical demonstration of oxidative damage in Alzheimer disease by immunochemical enhancement of the carbonyl reaction with 2,4-DNPH. J Histochem Cytochem 46:731–735

Sochor J, Ruttkay-Nedecky B, Babula P, Adam V, Hubalek J, Kizek R (2012) Automation of methods for determination of lipid peroxidation, lipid peroxidation. In: Catala A (ed). ISBN: 978-953-51-0716-3, InTech, 2012. https://doi.org/10.5772/45945

Spiteller G (2003) The relationship between cell wall, lipid peroxidation, proliferation, senescence and cell death. Physiol Plant 119:05–18

Spiteller P, Spiteller G (1997) 9-hydroxy-10,12-octadecadienoic acid (9-HODE) and 13-hydroxy-911-octadecadienoic (13 HODE); excellent marker for lipid peroxidation. Chem Phys Lipids 89:131–139

Spreitzer H, Schmidt J, Spiteller G (1989) Comparative analysis of fatty acid fraction of vegetables in dependence on preliminary treatment. Fett Wiss Technol 91:108–113

Stadtman ER (1992) Protein oxidation and aging. Science 257(5074):1220–1224

Stohs SJ, Bagchi D (1995) Oxidative mechanisms in the toxicity of metal ions. Free Radic Biol Med 18:321–336

Taylor NL, Heazlewood JL, Day DA, Millar AH (2005) Differential impact of environmental stresses on the pea mitochondrial proteome. Mol Cell Proteomics 4:1122–1133

Thomas DW, Van Kuizk FJGM, Dreatz EA, Stephens RJ (1991) Quantitative determination of hydroxyl fatty acids as an indicator of lipid peroxidation: gas chromatography and mass spectrometry methods. Anal Biochem 198:104–111

Thompson JE, Legge RE, Barber RF (1987) Role of free radicals in senescence and wounding. New Phytol 105:313–344

Tian YZ, Li WH, Shi GL, Feng YC, Wang YQ (2013) Relationships between PAHs and PCBs, and quantitative source apportionment of PAHs toxicity in sediments from Fenhe reservoir and watershed. J Hazard Mater 248(249):89–96

Timperio AM, Egidi MG, Zolla L (2008) Proteomics applied on plant abiotic stresses: role of heat shock proteins (HSP). J Proteome 71:391–411

Uchida K (2003) 4-Hydroxy-2-nonenal: a product of mediator of oxidative stress. Prog Lipid Res 42:318–343

Wang T, Powell WS (1991) Increased levels of monohydroxy metabolites of arachidonic acid and linoleic acid in LDL and arota from atherosclerotic rabbits. Biochem Biophys Acta 1084:129–138

Winston GW (1990) Physicochemical basis of free radical formation in cells: production and defenses. In: Smallwood W (ed) Stress responses in plants: adaptation and acclimation mechanisms. Willey Liss Inc, New York, pp 57–86

Yamauchi Y, Furutera A, Seki K, Toyoda Y, Tanaka K, Sugimoto Y (2008) Malondialdehyde generated from peroxidized linolenic acid causes protein modification in heat-stressed plants. Plant Physiol Biochem 46:786–793

Ying J, Clavreul N, Sethuraman M, Adachi T, Cohen RA (2007) Thiol oxidation in signaling and response to stress: detection and quantification of physiological and pathophysiological thiol modifications. Free Radic Biol Med 43:1099–1108

ROS and Regulation of Photosynthesis 5

Abstract

In plant, cell chloroplast is one of the prime locales for the formation of ROS and the origin of redox signal. Any redox imbalance in photosynthetic electron transport and photosynthetic carbon reduction cycle eventually causes generation of ROS in plants. An efficient antioxidative defense operates both at metabolic interface and at genetic level for processing ROS efficiently for the maintenance of redox homeostasis and ROS pool. The significance of antioxidative defense network in the maintenance of optimum photosynthetic rate has been revealed in many studies involving molecular genetics and proteomic approaches. Recent studies have confirmed that the internal redox state of some important components of Z-scheme electron carriers (plastoquinone, cytochrome b_6f complex, etc.) affects chloroplast gene expression, hinting the significance of chloroplast redox signal in controlling photosynthesis. Additionally, through redox regulation, photosynthesis functions as sensors for environmental cues like excess photochemical energy. This in fact provides regulatory loop in which expression of photosynthetic genes is not only coupled with redox state of photosynthetic electron flow but also senses excess photochemical energy. So, keeping all these views under consideration, an effort has been made to describe the present concepts of the role of photosynthesis in the origin of oxidative stress and redox signaling.

Keywords

Oxidative stress · Photosynthesis · Z-scheme regulation · Photooxidative damage · Redox signal

S. Bhattacharjee, *Reactive Oxygen Species in Plant Biology*,
https://doi.org/10.1007/978-81-322-3941-3_5

5.1 Introduction

A vast array of ROS generated during photosynthetic electron transport (PET) involving Z-scheme are formed as a result of single electron transfer to molecular oxygen principally during the Mehler reactions (Asada 1999). In case of PET, photoreduction of O_2 to superoxide takes place during pseudocyclic electron flow and commonly known as Mehler reaction (referred to Fig. 1.3), in honor of the discoverer (Mehler 1951). Even though photoreduction of molecular oxygen during pseudocyclic electron flow involving Mehler reaction is a vital option or alternative sink for the expenditure of surplus photochemical energy, it is constantly associated with the production of ROS and inclined to change the redox status of the organelle (Varnova et al. 2002). If the turnover of ROS becomes exceedingly higher, i.e., the formation of ROS exceeds the capacity of antioxidant systems to detoxify them, photodynamic injury to the photosynthetic apparatus results, which leads to significant loss of photosynthetic ability. The loss of redox homeostasis in terms of the acute dearth of $NADP^+$ in PSI [primarily due to comparatively slower pace of photosynthetic carbon reduction cycle (PCRC)] causes passaging of electron onto O_2 triggering the generation of $O_2^{\cdot -}$. Therefore, the regulated activation of PCRC and control pace of electron flow in Z-scheme of photosynthesis are significant factors that decide the redox state and fate of plant cell. This is enormously significant as the electron carriers of PSI, chiefly the ferredoxin, have adequate negative electrochemical potentials to provide electron to molecular O_2, ensuing superoxide ($O_2^{\cdot -}$) formation. The bulk of $O_2^{\cdot -}$ in vivo is considered to be generated via electron spilling from reduced ferredoxin to oxygen. Superoxide generated then undergoes dismutation either by the chloroplastic enzyme superoxide dismutase (Cu/ZnSOD) or even spontaneously. Further, it has been proposed that photoreduction of O_2 that leads to the origin of ROS in chloroplast ultimately detoxified to water by ascorbate-glutathione, glutaredoxin-thioredoxin, or Mehler ascorbate-peroxidase/catalase pathways under excess photochemical energy, which may involve about 30% of total PET (Bartoli et al. 1999). This also put forward the fact that O_2 is an alternative electron acceptor or sink in preventing photooxidative damages or photoprotection. As the formation of huge quantity of ROS is an expected outcome or inevitable consequences under excess photochemical energy (EPE), plants evolved different well-organized strategies by improvising different antioxidative defence mechanisms and integrating them with normal Z-scheme PET to overcome the imposed secondary oxidative stress (referred to Fig. 1.3).

The ROS singlet oxygen ($^1O_2$) is also incessantly generated during operational Z-scheme of photosynthesis under EPE involving mainly PSII. The PSII reaction center, a heterodimer of D1 and D2 proteins, apart from associating with cytochrome b_{559}, also enables the binding of functional prosthetic groups of photosynthetic pigments and redox components of PSII. Under EPE the redox state of plastoquinone pool and Q_A and Q_B are over-reduced which may cause oxidized P_{680} to recombine with reduced pheophytin and favor the generation of triplet state of P_{680}, leading to the generation of singlet oxygen (electron of outermost orbital start rotating in reverse direction) by subsequent resonance transfer to molecular oxygen. It is found that EPE causes noteworthy augmentation in generation of singlet oxygen which subsequently causes photoinhibition (Hideg et al. 1998, 2002).

Photosynthetic electron transfer (PET)-mediated ROS production is usually explained as injurious symptom because the overproduction always causes oxidative damage to thylakoid membrane and other important components of chloroplasts, eventually blocking photosynthesis, what is commonly called as photooxidative damage. So, there was an effect for the plant cell to minimize ROS production by integrating Z-scheme of electron flow with antioxidative defense circuit. However, despite their harmful effect, it is currently well recognized that ROS are also potent-regulating molecules that are concerned in the regulation of plant growth and development and also to a large extent determined plant acclimatory responses under stress (Foyer and Noctor 2009; Bhattacharjee 2012; Foyer 2015). In fact, the signaling function of the powerful oxidants (ROS) is largely ignored, and the ROS functionality is largely biased solely on the notion that ROS exert the principal effects through chemical toxicity and oxidizing ability. So the physiological implication of the "oxidative stress" is largely influenced by nonspecific "photooxidative damage" to cellular components, where the oxidative deterioration goes beyond of repair under a situation of disruption of redox homeostasis. Furthermore, it is often hypothesized that this kind of nonspecific oxidative damage to important cellular macromolecules may initiate necrosis or cell death.

Recent literature revealed that photosynthetic generation of ROS under the influence of different environmental conditions, particularly excess photochemical energy, has serious implication in signaling and plant performance under stress, apart from their toxic role (Foyer and Shigeoke 2011; Bhattacharjee 2014; Foyer and Noctor 2009). In fact the internal redox cue of chloroplast offers necessary information to the cell about the metabolic balance between energy-generating and energy-utilizing process. In fact the redox imbalance in chloroplast associated with higher pace of Z-scheme of electron flow is considered as powerful signal that not only decreases PSII activity but also regulates the expression of defense gene necessary for plant performance under stress. So the choice between "oxidative damage" and "oxidative signaling" in data analysis associated with redox status of chloroplast became extremely crucial for the evaluation of proper physiological status of ROS and oxidative stress in plants.

5.2 Origin of Photooxidative Stress

In plants, dissipation of EPE is an instantaneous and delicately tuned response which occurs through a number of mechanisms, which include heat irradiation, emission of light having higher wavelengths and dispatching photosynthetic electrons (from excited reaction center after resonance transfer of energy from antenna pigments) to alternative sinks, and ultimately downregulation of photosystem II (Foyer 1996; Huner et al. 1998). The incidence of photoreduction of O_2 is an optional sink for photosynthetic electron in PET chain (under pseudocyclic electron flow), especially when there is a significant deficiency of $NADP^+$. But this spilling of electron to molecular O_2 is always associated with the formation of ROS, such as $O_2^{\cdot-}$, H_2O_2, $OH^{\cdot}$, etc. (Alscher and Hess 1993; Foyer 1996). Moreover, over-excitation of PSI can also dissipate their energy by transferring the energy to O_2, causing reversal in electron movement and generation of ROS $^1O_2$. If accumulation of ROS under conditions of EPE exceeds the capacity of enzymatic and nonenzymatic antioxidant systems to scavenge them, then photooxidative damage to photosynthetic apparatus ensues (Mehler reaction), which leads not only to blockage of photosynthesis but also cell death. This is often manifested by the appearance of chlorotic lesions on photooxidatively damaged leaves.

In contrary to photooxidative stress and damage, ROS may also take part in significant positive role in response to EPE by initiating an increase in rate of degradation of DI protein of LHCII (Fig. 5.1). This causes physical separation of reaction center of PSII and its LHCPs, causing photoinhibition of photosynthetic electron flow, which may be considered as a photo-protective mechanism under EPE. The prospect of both positive and negative roles for ROS or induced redox changes under different magnitude of photooxidative stress in chloroplast experiencing EPE suggests that it is more appropriate to view ROS turnover as a significant mechanism for regulation of photosynthesis. In other word, the changed equilibrium between the processes that produce ROS and the efficiency of combating antioxidative defense which reduce them rather than the levels of these antagonists perse are significant in determining the fate of photo-oxidative stressed tissue (Foyer 1996). However, the instantaneous responses to excess light energy may lead to a whole plant acclimation through an amendment to the photosynthetic competence of new leaves in which ROS have significant function. Previous works categorically

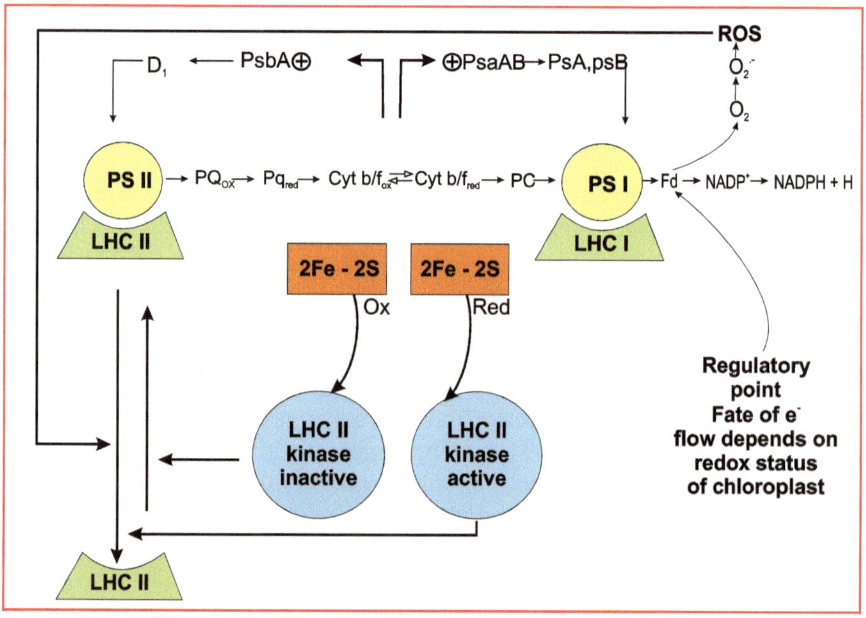

Fig. 5.1 Redox-regulation of photo-inhibition and the influence of redox state of Fe cytochrome b/f system of Rieske center on the expression of photosynthetic genes (PsA, PsB). (Details in the text)

emphasize the role and impact of oxidative stress in response to EPE on acclamatory performances of plants (Foyer et al. 1997; Krause 1988; Russell et al. 1995). When a leaf experiences EPE, the instantaneous response is changes in internal redox cues and signaling in the proximity of PSII, causing differential expression of chloroplastic antioxidant defense genes. However, extended experience under EPE leads to the death of chloroplastic cells of such leaves. In fact, leaves exposed to EPE are capable of generating a systemic redox signal, a component of which may be H_2O_2, or even the redox status of some important PET chain carrier, which ultimately instigate a downstream signaling event causing stress acclamatory response in unstressed regions of the plant. This redox signaling leads to an improved ability to withstand further episodes of photooxidative stresses by the upregulation of antioxidant defenses.

One important experimental evidences revealed that EPE applied to low light (LL)-adopted *Arabidopsis* induces an oxidative burst that leads to the reversible photoinhibition (Santos et al. 1996). Astonishingly, it was observed that such chloroplastic oxidative stress or redox cues only upregulate genes encoding key components of cytosolic ROS scavenging systems (APX2 or ascorbate peroxidase isoform), through retrograde signaling. The differential expression of this nuclear antioxidative gene is found to regulate significantly rapid changes in the redox status of plastoquinone pool (PQ) and the activity of associated PSII. Further evidences of redox regulation of gene expression come from the experiment with glutathione, which was found to block the induction of APX2 by EPE, implying that changes in the cellular redox pool might have a key role in retrograde redox signaling (Noctor et al. 1998).

Exogenous treatment of leaves with H_2O_2 followed by exposure to excess light (EL) caused significantly better stress acclimation through upregulation of APX2 as compared to control EL alone. This interesting observation, when investigated in a sequence of time-course experiments, revealed the fact that leaves pretreated with H_2O_2 showed a slower decline in maximal PSII efficiency under EL than control leaves, hinting at the role of prooxidant status in adapting excess light energy. Similarly, the protective effects of ROS (H_2O_2) in triggering various stress acclimation mechanisms have been explained for chilling-stressed maize seedlings (Krause 1988; Foyer 1997).

Though all those evidences do not implicate whether H_2O_2 has a direct or indirect role, they strongly lay emphasis on the role of this photosynthetic ROS in the stress acclimation, particularly to conditions of EPE. The just conflicting effects of aggravated PSII inhibition upon treatment with reduced glutathione, which reduces H_2O_2 by enzymatic and nonenzymatic reactions, are consistent with such function of H_2O_2. Thus ROS can significantly influence the turn over of glutathione i.e. recycling of GSSG and GSH by modifying thiol/disulphide ratio, thereby influencing on redox status of the cell which ultimately can bring about photosynthetic redox signaling (Foyer et al. 1997) (Fig. 5.2).

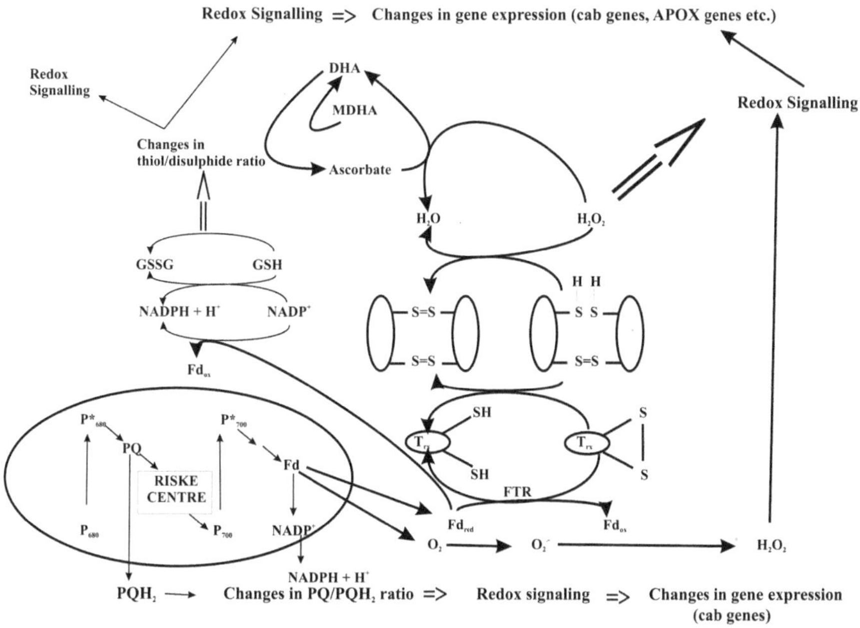

Fig. 5.2 Z-scheme-associated origin of redox signals and their consequences leading to gene expression. (Detail in text)

5.3 ROS-Antioxidant Interaction and Regulation of Redox Status of Chloroplast

The light-driven photosynthetic electron transfer (PET) in thylakoid membrane cleaves electrons from H_2O to NADP+ through redox carriers generating H^+ gradient that facilitates ATP synthesis. However, one of the key features of PET is its susceptibility to EPE-induced redox imbalance and subsequent oxidative damage (Fig. 5.1). Capture of excess photochemical energy (EPE) instigates the faster PEF and greater utilization of NADP+ than its regeneration under PCRC, ultimately causing significant deficiency of the terminal electron receptor in Z-scheme. As a result, the further excitation of LHCPs by EPE and subsequent PEF causes passage of electron to molecular O2 (pseudocyclic electron flow) causing incomplete reduction of O_2 and generation of $O_2^{\cdot -}$ (Mehler reaction). Formation of $O_2^{\cdot -}$ subsequently caused the subsequent generation of other ROS like H2O2, OH, etc. Similarly, over-energization of PSII by EPE often transfers the energy to O_2, reversing the direction of electron flow and the genesis of $^1O_2$. So, a significant feature of PET, in particular, under EPE, is its vulnerability to light-induced production of ROS and associated photooxidative damage to chloroplast (Vass and Cser 2009; Foyer and Noctor 2009). One of the consequences of such damage is irreversible oxidation of the D1 protein of PSII reaction center (Kriger-Liszkay et al. 2008). It further necessitates the reorganization of PSII reaction center in order to overcome the light stress. But if the disorganization of PSII exceeds the rate of its reorganization, it causes photoinhibition (Vass and Aro 2008). But in many instances, it is found that the rate of repair of D1 protein of PSII is as high as the rate of damage, maintaining an elevated level of PSII action (Ohad et al. 1984). So, plant needs to have a competent regulation of PEF to reduce the formation of $^1O_2$ at PSII and $O_2.^-$, H_2O_2, OH· at PSI (Genty and Harbinson 1996). Reversible reduction in the efficiencies of both the photosystems is intrinsic to the regulation of utilization of EPE during photosynthesis (Genty and Harbinson 1996). In fact, the inherent limitations of capacities of electron flow through the cytochrome b6/f complex of Rieske center always favor over-reduction of PSII that exacerbate PSII turnover (Vass and Cser 2009). Another important strategy that is generally used to defend PSII under EPE and low PET ability is non-photochemical quenching (NPQ) that dissipates excitation energy of antenna molecules as heat energy (Niyogi et al. 1998, Foyer and Shigeoka 2011).

Operation of cyclic electron flow may be another strategy for the prevention of over-reduction of acceptor side of PSI, whereby ATP production increased relative to NADPH++H+ (Munekage et al. 2008). This is very significant because ROS are also produced at the acceptor side of PSI. Cyclic electron flow in general enhances protonation in thylakoid lumen and thereby limits the $^1O_2$ production that subsequently triggers NPQ machinery (Rumeau et al. 2007).

So, the precise role of noncyclic, cyclic, and pseudocyclic electron flow (PCEF) depends primarily on environmental condition. In C3 plants, non-cyclic (NCEF) and cyclic electron flow (CEF) are generally governed by relative demand for ATP and NADPH+H+, where as in case of pseudocyclic electron flow (PCEF) PSI is considered to be the key site for the formation of ROS, depending on the redox status and environmental condition. For the augmentation of ATP production relative to NADPH+H+, where there is a balance demand for both NADPH+H+ and ATP, noncyclic electron flow (NCEF) is instigated. On the contrary PCEF, to a large extent, depends on environmental condition, particularly EPE or light stress.

While many basic fundamental questions concern the NCEF, CEF, and PCEF and the degree to which this pathway functions in C3 plants, there is a common agreement that although NCEF ideally feeds the plant with both fuels necessary for photosynthesis, CEF is considered as essential component of repertoire of chloroplastic photosynthetic mechanism that serves to coordinate energy metabolism with redox status. Another important mechanism of regulation of photosynthetic electron flow is through prevention of over reduction of PSI acceptor and chloroplast stroma is the 'match valve' system that transfer reducing equivalent to cytosol for the regeneration of $NADP^+$ (Scheise et al. 2005). This pathway employs thioredoxin (TRX)-mediated upregulation of chloroplastic $NADP^+$-dependent malate dehydrogenase and is considered as vital part for the regulation of stromal redox homeostasis, as it allows the export of surplus reducing power and regeneration of $NADP^+$, thus mitigating pseudocyclic electron flow and relieving electron pressure in chloroplast. So the regulation of distribution of metabolites may be used by green cell to restore redox homeostasis in order to achieve photosynthetic control. However the contribution of antioxidant network in the regulation of distribution of metabolites and in the maintenance of stromal redox homeostasis is another fascinating area of the regulation of photosynthesis.

5.4 An Elaborate Antioxidant Network Works at Chloroplast for the Management of Oxidative Stress and the Origin of Redox Signal for Combating Oxidative Stress

Being an important source of cellular oxidants, even under normal condition, the estimated rate of formation of ROS in chloroplastic PET as Z-scheme in C3 plants is nearly 4000 nmolm^{-2} s^{-1}, in contrast to peroxisomal rate which is 10000 umolm^{-2} s^{-1} (Foyer and Noctor 2003). This magnitude of ROS formation, however, got enhanced several times under environmental stress such as EPE and dehydration although it is very hard to assess quantitatively the exact titer of ROS like H_2O_2 in chloroplast because of inherent complexities of the system (Queval et al. 2008). However regardless of uncertainties of ROS production in chloroplast, it has long been documented that ROS and H_2O_2 in particular are effective inhibitors of photosynthesis. Even at very low titer of H_2O_2 (10uM), treatment can inhibit 50% of CO_2 fixation, primarily because of oxidation suffered by thiol-containing and thiol-modulated enzymes. So the redox balance involving the rate of generation of ROS and their scavenging in chloroplast is extremely important and needs to be tightly regulated. The following mechanisms offer tight regulation of ROS accumulation of chloroplast:

5.4.1 Ascorbate-Glutathione (ASA-GSH) Cycle or Halliwell-Asada Pathway

Although both ascorbate and glutathione, the water-soluble antioxidants, got the capacity of scavenging ROS separately, they best perform their antioxidant role in ASA-GSH cycle (Foyer and Halliwell 1976; Foyer and Shigeoka 2011) and water-water cycle to metabolize H_2O_2 (Asada 1999) to dissipate excess H_2O_2 (Fig. 1.3).

The redox imbalance, caused primarily by deficiency of $NADP^+$ during EPE, causes reduction of ground-state molecular O_2 from over-reduced on the acceptor side of PSI, in the so-called Mehler reaction. This step of monoelectronic reduction of O_2 by reduced ferredoxin in PSI is the starting point of "water-water cycle" (Asada 1999). This cycle further incorporates SOD, APOX, and other ASA-GSH cycle components (Fig. 1.3) in a mechanism that builds up transthylakoid proton gradient and facilitates ATP generation at the expense of NADPH (Asada 1999). APOX utilizes ASA as electron donor to reduce H_2O_2 to H_2O with concomitant production of monodehydroascorbate (MDA). MDA in turn is spontaneously converted to ascorbate (ASA) and dehydroascorbate (DHA) by the action of NADPH-dependent MDA reductase. DHAR utilizes GSH to reduce DHA and thereby regenerate ASA. Oxidized glutathione (GSSG) produced thereby subsequently reduced to GSH by glutathione reductase (GR) using $NADPH+H^+$. Previous work reported the presence of different APx and SOD isoforms in stroma and thylakoid membranes; the chloroplastic GR and DHAR are found to be localized in stroma. One of the very important properties of chloroplastic APx is their susceptibility toward oxidative inactivation, particularly in the absence of ASA (Ishikawa and Shigeoka 2008).

Depletion of chloroplastic ASA level and subsequent inactivation of APxs, therefore considered as prime factor regulating the efficiency of photosynthesis, under stress, may be the potential target of crop improvement. Peroxiredoxin (PRx and PrxQ), glutaredoxin (GRx), glutathione peroxides, sulfiredoxin, and cycloplins are other classes of proteins which work in tandem with TRx-like proteins in chloroplastic detoxification of ROS (Dietz 2011). All these thiol-based detoxification mechanisms used by PRx to reduce ROS, particularly H_2O_2, mainly involve peroxidative reduction, followed by regeneration involving several electron donors such as glutaredoxin, glutathione, ascorbate, etc. Glutathione peroxidases may exploit both GSH and TRx as reductants to detoxify H_2O_2 and lipid peroxides produced during lipid peroxidation. TRx in general have been found to be more efficient in detoxifying H_2O_2 compared to GSH (Dietz et al. 2002). In a GPX-mediated detoxification process, regeneration of TRx is linked to Z-scheme of electron flow involving ultimately Fd-TRx reductases, a NADPH-dependent TRX reductase. PRx and GPx pathway therefore provide alternative pathway to water-water cycle under excess photochemical energy primarily because significantly of lower affinity of GPXs toward H_2O_2 than APx and PRX. Lipid peroxides produced by membrane lipid peroxidation are also efficient substrate for chloroplast GPxs and recognized as extremely important event in chloroplast for the maintenance of redox homeostasis and maintenance of optimum photosynthetic rate under stress. Regulatory compensatory mechanisms also exist between the ASC-GSH- and PRX-dependent pathways. It was observed that chloroplastic APx and MDA reductases are enhanced in antisense *Arabidopsis*, plants with suppressed chloroplastic PRx activity (Baier et al. 2000). Such types of observation strongly favor the existence of cross talk between ROS detoxification pathways in chloroplasts.

So, the detoxification of ROS generated as a natural by-product of Z-scheme of electron flow always depends on ASC-GSH cycle or TRx-dependent pathway within stroma. In fact the efficiency of ASC-GSH pathway to a larger extent depends on the function of a-tocopherol and carotenoid hydrophobic antioxidants. A-tocopherol is found to inhibit chloroplastic lipid peroxidation by scavenging $^1O_2$ generated in Z-scheme under stress (DellaPenna and Pogser 2006; Kriger-Liszkey et al. 2008). The experimental evidence in support of the existence of close relationship between chloroplastic antioxidant systems may be vouched by studying the relative level of antioxidant efficiency in *Arabidopsis* mutant.

Arabidopsis Vte1 mutant that is deficient in a-tocopherol shows significant accumulation of ASC and GSH. On the other hand, plants overexpressing vit. E have lower accumulation of GSH and ASC (Karwischer et al. 2005). So, these kinds of experimental evidence not only suggest the close relationship between AOX pathways but also exhibit reciprocal regulation of AOX pathway in chloroplast for the maintenance of redox homeostasis (Smirnoff et al. 2001; Foyer and Shigeoka 2011). Therefore, the factors that manage the intracellular separation of metabolites and AOX between the different cellular compartments are extremely important for overall regulation of photosynthesis particularly under stress. Although in recent years understanding of ROS signaling that coordinates expression of AOX gene in chloroplast has increased significantly, little information is available till date on the role of low molecular weight antioxidants in the process. EPE often lead to over-excitation of LHCPs that often results in photoinhibition of Z-scheme electron flow. The most significant impact of the EPE is oxidative charge of D1 protein in PSII (Genty and Harbinson 1996). To avoid the EPE-mediated photoinhibition, other quenching processes that dissipate the excess energy are via heat or Mehler valve.

5.4.2 Water-Water Cycle or Asada Pathway

The dissipation of EPE absorbed by the photosynthetic apparatus and its management is an essential process for the survival and growth performance of almost all photosynthetic organisms. An efficient management of EPE prevents photooxidative damage that occurs when excited chlorophyll molecules improperly transfer their higher-energy state to molecular oxygen, starting genesis of ROS. This process in particular is extremely crucial when CO_2 fixation by PCRC is limited because of environmental stress (Alsher and Hess 1993). Under such conditions, the energy absorbed by the chloroplasts cannot be funnelled for the reduction of CO_2, causing serious loss of balance between PET and PCRC. Maintenance of PET through the photosynthetic membrane, even under extreme unfavorable environmental conditions, is therefore vital for preventing oxidative injury to plant cells. Out of the different adaptive strategies that are thought to cooperate in protecting the photosynthetic apparatus from photooxidative damages, zeaxanthin cycle directly protects the LHCP-bound antenna molecules and the other options like cyclic electron flow and the water-water cycle (Asada pathway) that shunt electrons through the photosynthetic Z-scheme and maintain the pH gradient in the chloroplast.

In water-water cycle, electrons obtained from the photolysis of water involving OEC (oxygen-evolving complex) at PSII are ultimately transferred to oxygen through PSI and result in the formation of superoxide radicals and then to H_2O_2 by Cu/ZnSOD. Eventually, a thylakoid membrane-bound enzyme thylakoid-APX converts H_2O_2 back into H_2O. Ascorbic acid, used by the thylakoid-bound APX as a reductant, is converted during this process into ascorbic acid radical (monodehydroascorbate), and this radical is reduced back to ascorbic acid by ferredoxin using electrons from PSI or by reduced glutathione. Therefore, the physiological significance of the water-water cycle or Asada pathway is to maintain the incessant electron flow through Z-scheme PET, even when CO_2 fixation is limited or PCRC is significantly inhibited (Noctor and Foyer 1998). Essentially, it completes the cyclic flow of electrons from one water molecule associated with PSII to another water molecule, i.e., the product of a different antioxidative reaction, which is again at close proximity to another photosystem, i.e., PSI. Thus, it maintains the consequence of PET, i.e., the proton pumping across the thylakoid membrane through plastoquinone and OEC, necessary for the generation of transmembrane proton gradient and sustains the process of photophosphorylation.

Even though the significance of the water-water cycle in the management of EPE was revealed from a number of different physiological and metabolic studies, molecular genetics evidences corroborating the photo-protective role of this pathway is very limited. Some relevant studies have shown that augmenting the expression of the thylakoid-attached Cu/ZnSOD or the thylakoid-bound APX improves the abiotic stress (particularly light stress) tolerance of transgenic plants. On the other hand, the molecular genetics experiments involving mutants with loss of function studies for these two enzymes were not published, and the physiological significance of the water-water cycle in maintaining chloroplast physiology in the absence of unfavorable environmental stresses was not documented. It is primarily because the loss of function of these enzymes happened to be lethal. Some workers also question the function of the water-water cycle in defending the oxidative damage of photosynthetic apparatus by asserting that the quantum of electrons transferred through this pathway is very little and not adequate to protect the chloroplast from photooxidative stress. Recently some workers with molecular genetics approaches knockdown *Arabidopsis* plants with downregulation of thylakoid-attached Cu/ZnSOD to reveal that the water-water cycle is necessary for the defense of chloroplasts both in the presence and absence of environmental stress conditions that limit the availability of the electron acceptor $NADP^+$ in chloroplasts, under redox imbalance state (Foyer and Shigeoka 2011).

5.4.3 ROS-Induced Photoinhibition and Its Significance in the Maintenance of Redox Homeostasis of Green Cell Under Stress

Excess photochemical energy (EPE) can cause redox imbalance and cause oxidative stress by producing excess ROS as inevitable consequence of decontrolled PET (refer to Fig. 1.3). Photoreduction of molecular oxygen on the acceptor side of photosystem I (PSI), in Mehler reaction, as a result of the Z-scheme electron flow or photosynthetic electron transfer (PET), leads to the generation of a number of ROS sequentially, like $O_2^{\cdot-}$, H_2O_2, and $OH^{\cdot}$. Likewise, another outcome of EPE is shift of excitation energy from excited antenna pigments and reaction center associated with LHCPs of PET to molecular oxygen leading to the formation of $^1O_2$. The endogenous titer of ROS can be reduced to significantly tolerable range by antioxidative systems that include ROS scavenging systems, such as ascorbate-glutathione cycle, glutaredoxin pathway, and activities of other antioxidative enzymes superoxide dismutase and ascorbate peroxidase, as well as low molecular weight antioxidants and quenchers, such as glutathione, ascorbate, vitamin A and E, etc. However, when the absorption of light exceeds the capacity of PCRC or the photosynthetic machinery for photosynthesis, i.e., EPE, the production of ROS is an inevitable consequence, and elevated levels of ROS cause loss of redox homeostasis or oxidative stress and damages.

Participation of ROS in the photooxidative damage to PSII has also been a subject of argument for a long time. A number of investigations suggested that H_2O_2 and $^1O_2$ are the primary cause of photooxidative damage to the components of PSII. Mechanisms, like the "acceptor-side" and "charge-recombination" systems, are proposed for ROS-mediated oxidative damage. Under EPE, $^1O_2$ can be noticed significantly in isolated PSII complexes or in isolated thylakoid membranes and in plant leaves. Under this situation, exposure of $^1O_2$ to thylakoid membranes results in specific cleavage of the D1 protein of LHCII complexes. Other ROS, such as $O_2^{\cdot-}$, H_2O_2, and $OH^{\cdot}$, also induce the specific cleavage of the D1 protein of PSII in vitro. On the other hand, there are other mechanisms of photooxidative damage to PSII which may not be caused by ROS. In the alternate proposed "donor-side" mechanism, excessive acidification of the thylakoid lumen, due to the transfer of protons across the thylakoid membrane associated with PET, inactivates the OEC (oxygen-evolving complex) and allows the oxidized reaction center ($P680^+$) to survive for an extended period of time with strong oxidizing capability. This strong oxidant ($P680^+$) often damages the D1 protein, causing photoinhibition (Krause 1994). In vitro studies involving isolated PSII and thylakoid membranes also support this alternate mechanism making the entire concept doubtful in the photoinhibition of PSII. Therefore, there may be different methods for controlling photooxidative damage under in vivo system in plants. Apart from this, the photooxidative damage to PSII is light-dependent and may not be mediated by ROS. Therefore, a new idea for the role of ROS in photooxidative damage to PSII may be proposed where the role of prooxidants in the repair of PSII is suggested. In fact, protein synthesis is found to be precise target of ROS under such condition. In fact, photooxidative-damaged PSII is restored in several steps, like:

1. Proteolysis of the D1 protein.
2. De novo synthesis of the precursor to D1 protein.
3. Inclusion of the nascent pre-D1 peptide into the thylakoid membrane and its assembly with other components of damaged PSII.
4. Processing of carboxy-terminal of pre-D1 peptide for the maturation of D1 protein.
5. The functional assembly of OEC (catalytic oxygen-evolving complex).

In order to identify the site of inhibition of repair by these ROS ($^1O_2$, hydrogen peroxide) in the sequence of steps that lead to the de novo synthesis of the D1 protein in algal cell, all the steps like transcription of the psbA gene for D1, translation of psbA mRNA, and processing of pre-D1 were monitored. Northern and immunoblotting analyses revealed that the translation of psbA mRNA is downregulated by ROS. Experimental evidences also suggested that the target of ROS might be the elongation step of translation. So, the suppression of translational elongation of psbA mRNA under oxidative stress might cause the inhibition of the repair of PSII during the photoinhibition (Fig. 5.1). It is also found that the photooxidative damage to PSII occurs primarily by UV and low wavelength blue light involving OEC; on the contrary, the secondary injury takes place by visible light absorbed by antenna pigments associated with the reaction center of PSII. The release of Mn^{2+} from the fragile OEC is mediated by oxidative damage to PSII, signifying the likely disorganization of the manganese cluster upon perception of damaging radiations. Further, once the OEC is damaged, the contribution of electrons by photolysis of water to the reaction center of PSII (P_{680}^+) for regeneration of P_{680} ground state is blocked causing its accumulation, which in turn damages the reaction center by oxidizing the D1 protein.

5.4.4 Xanthophyll Cycle and Its Role in Photoprotection

Over-excitation of LHC complexes due to excess photochemical energy leads to the overutilization and $NADP^+$ pool, causing serious dearth of the electron acceptor, thereby compelling pseudocyclic electron flow and production of ROS. Not only that, over-excitation of LHC complexes and corresponding photosystems always exhibited an enhanced lifespan of the antenna pigments, thus furthering the possibility of formation of chlorophyll triplet state, which may react with oxygen, causing reversal of movement of electron flow in the outermost orbital i.e. genesis of singlet oxygen, another ROS. Excess generation of ROS, beyond the capacity of integrated antioxidative defense system (ASC-GSH pathway, glutaredoxin pathway, etc.), always leads to loss of redox homeostasis and evokes oxidative stress in chloroplast.

To combat this situation, some of the products of the light-triggered xanthophyll cycles take part in a significant role in the defense against the loss of redox homeostasis and oxidative damages caused by EPE and other forms of environmental stresses. In this connection, at least four feasible mechanisms of xanthophyll cycles for conferring protection against oxidative stress of chloroplasts may be considered:

1. The first one involves the direct quenching of over-excitation from excited antenna pigments by products of the light phase of xanthophyll cycles.
2. The second option is the indirect participation of xanthophyll cycle. Here the carotenoids can indirectly quench over-energized pigments through aggregation-dependent LHCII quenching.
3. The third option involves the light-driven mechanisms in LHCII.
4. And finally the charge transfer quenching mechanism between Chla and Zx.

Light absorption during Z-scheme of photosynthesis is basically accomplished by antenna pigments associated with LHCPs which are intricately linked with reaction centers. Absorption of photon by Chl (as antenna pigment) generates singlet-state excitation of a Chl molecule ($^1Chl^*$), which may subsequently return to the ground state by means of one of the various options (Fig. 5.3). First of all, the excitation energy of $^1Chl^*$ can be released as Chl fluorescence (emission of light with higher wavelengths), it can transfer the energy as resonance to the reaction centers (P-680 or P-700) and used to drive PET and generation of reductant and metabolic energy needed for PCRC, or even it can be de-excited by thermal dissipation processes, i.e., non-photochemical quenching (NPQ). Sometimes the energy from $^1Chl^*$ can decay via $^3Chl^*$ (the triplet state) alone. Although the triplet pathway can be a significant valve for the expenditure of excess excitation energy, the $^3Chl^*$ can transfer energy to ground-state O_2 to generate singlet oxygen ($^1O_2^*$), an extremely

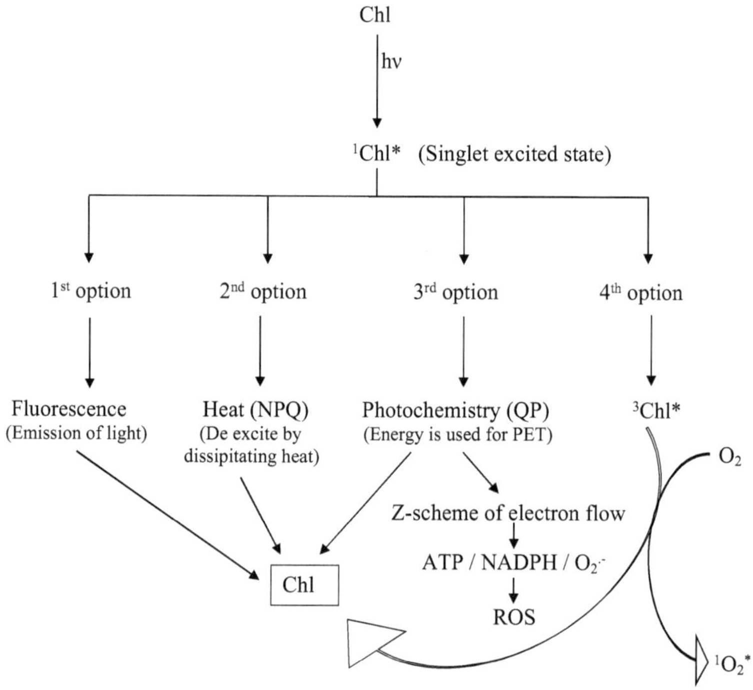

Fig. 5.3 Chlorophyll excitation and its subsequent consequences during photosynthesis. (Detail in text)

damaging reactive oxygen species (Foyer and Harbinson 1999). The yields of ^{3}Chl* and fluorescence vary in proportion to the standard lifetime of ^{1}Chl*, which in turn depends on the efficiency of other pathways. For example, the high quantum efficiency of photochemistry in limiting light results in a decrease or quenching of fluorescence that is termed photochemical quenching (qP). Other non-photochemical processes that disperse and dissipate excitation energy also quench Chl fluorescence and are collectively called NPQ (or qN) (Bilger and Björkman 1994; Gilmore 1997).

The decrease in lumen pH as a consequence of upregulated Z-scheme of photosynthesis under EPE activates the interconversion of specific xanthophyll pigments (oxygenated carotenoids) that are mostly bound to LHCPs. This interconversion occurs on a timescale of minutes as part of a xanthophyll cycle. All organisms that exhibit qN have a xanthophyll cycle, of which there are two main types. The violaxanthin cycle is associated with higher plants, green and brown algae. It basically involves the pH-dependent conversion from violaxanthin, a xanthophyll with two epoxide groups to zeaxanthin (no epoxide group) via antheraxanthin (one epoxide group). Eukaryotic algae have a different xanthophyll cycle (the diadinoxanthin cycle) that involves a conversion from diadinoxanthin, with one epoxide group, to diatoxanthin (without epoxide group) (Havaux et al. 2000; Niyogi et al. 2001).

In higher plants the de-epoxidation reaction is catalyzed by violaxanthin de-epoxidase (VDE), which is a 43-kD nucleus-encoded protein that is localized in the thylakoid lumen (Bugos and Yamamoto 1996). It is found that VDE enzyme is activated by low pH. Upon acidification of the lumen during active PET, VDE is found to be closely associated with the thylakoid membrane, where it can interact with its substrate violaxanthin (Hager and Holocher 1994). VDE subsequently exploits vitamin C to reduce the epoxide group, as evident from its from pH-dependent K_m values for ascorbic acid (Bratt et al. 1995).

Other enzyme, zeaxanthin epoxidase (ZE), catalyzes the epoxidation reactions that complete the violaxanthin cycle. ZE, which is a FAD-containing O_2-dependent mono-oxygenase, uses reduced ferredoxin ($FADH_2$) to epoxidize first zeaxanthin and then antheraxanthin (Bouvier et al. 1996). Because of its pH optimum 8, ZE is positioned on the stromal side of the thylakoid membrane to be constitutively active. The availability of zeaxanthin is therefore tightly determined by the activity of VDE as compared to ZE, with rapid accumulation of zeaxanthin occurring upon activation of VDE under EPE. Both ZE and VDE are the members of the lipocalin family, a diverse group of proteins that attach small lipophilic molecules and exhibit a conserved tertiary structure (Bugos et al. 1998; Tardy and Havaux 1997).

All of the xanthophyll cycles are engaged in antioxidant defense in plant cells and hence play vital role in redox regulation of cell. Products of the light-dependent phase of these cycles play an important role in the protection against oxidative stress generated not only by EPE but also by other ROS-generating factors like drought, chilling, heat, and salinity stress. It was demonstrated that these products are effective quenchers of ROS. Several molecular mechanisms exist to explain the photoprotective role of the xanthophyll cycle. Some of them are based on facilitation of the energy dissipation in the photosynthetic apparatus, which results in the decrease in production of ROS like singlet oxygen and other free radicals under over-excitation conditions. The others refer to direct quenching of ROS. The significant mechanisms capable of explaining the defensive role of xanthophyll cycles in photooxidative stress of higher plants are basically based on the indirect involvement of the de-epoxidized pigments, which include:

1. Aggregation-dependent LHCII quenching of over-excitation.
2. Mechanisms associated with light-induced changes of LHCII.
3. Charge transfer quenching between Chla and Zx.

5.5 Origin of Chloroplast Redox Signal and Control of Photosynthesis

So, EPE lead to an over-excitation of the light-harvesting complexes, which may cause photoinhibition by oxidative modification of D1 protein or suppression of the synthesis of the D1 protein in PSII. To keep away from this phenomenon, several quenching processes and antioxidative defense system dissipate EPE via heat or the Mehler valve. In contrast, under optimum light conditions, a light quality-dependent inequality is often noticed in the pattern of excitations of photosystems which might be neutralized by modifying the stoichiometry of PSI and PSII or by inducing a state transition. State transition is a short-term response, where the inequity in absorption of excitation energy between the PSII and PSI is counterbalanced by dissociation and the lateral movement of light-harvesting complex II antenna complexes across the thylakoid membrane (mobile part). On the contrary, adjustment of stoichiometry of photosystems is a long-term response in which the relative quantum of photosystems is altered in support of one which rate is limiting that ultimately cause enhancement in the competence of photosynthesis.

The plastoquinone (PQ) pool and more precisely its redox state, which revealed the poise of light use between both the photosystems, also control and regulate PET both under normal and stress acclamatory conditions. The light stress acclamatory performances are based on the changes in the redox state of photosynthetic redox components, particularly the PQ pool or the generation of ROS. Here, many of these redox-reactive regulatory molecules act as signals that can even modulate the expression of chloroplast and nuclear genes. All the genes that exhibited redox-regulated expression characteristics are directly associated with light trapping processes (Fig. 5.1).

It is widely accepted that the regulation of nuclear and chloroplastic genes is different. Nuclear genes are basically regulated at the transcriptional level whereas the chloroplastic genes at the posttranscriptional level. Nevertheless, in recent times, studies from many laboratories exhibited that majority of the steps in gene expression have been found to be very much regulated, both in the nucleus and in the chloroplast. In photosynthetic organisms, chloroplast redox signals that originated particularly under stress initiate important regulatory events that act on differential gene expression by a range of mechanisms (Fig. 5.1). Several works eventually showed that redox regulation of photosynthesis genes occurs at multiple expression levels with an extremely complex signaling network. The identification of chloroplast redox signal at proper level seems to be a difficult task. Contradictory or inconclusive results from different research groups, especially in the detection of the origin of chloroplast redox signals, either from the PQ pool or the cytochrome b_6f (Rieske center) complex, are evident (Fig. 5.2). Moreover, a specific control or regulation at a certain level of gene expression is not necessarily reflected on the next one. As a result, it is significant to analyze each step, i.e., right from transcription and transcript pool size to translation of redox-regulated genes in different plants to obtain an absolute picture of this and the types of regulation. In conclusion, it can be said that our understanding on redox genomics combined with proteomics and metabolomics might throw light on global information of molecular network of chloroplast redox signals.

5.6 Conclusion and Perspective

Chloroplast being one of the major sources of ROS regulatory systems is required not only to tightly regulate the turnover of ROS but also for the maintenance of redox status of the organelle, which directly governs the photosynthetic competence, growth, and development of plant. To maintain redox homeostasis and control generation and endogenous titer of ROS particularly under different types of environmental assaults (EPE), chloroplast operates a complicated redox signaling network which primarily senses stress-induced redox imbalances and initiates compensatory defense processes to maintain redox homeostasis that ultimately affords photoprotection. So, by making the photosynthetic ROS removal process efficient, particularly under unfavorable environmental conditions, photosynthesis remains unperturbed by environmental odds. In fact, the antioxidant efficiency is maintained and regulated (or buffers) in such level that never completely removes ROS but allows these prooxidants (signals) to persist within the chloroplast and outside for both signaling at the site (chloroplast) and evoking a retrograde signaling necessary for differential gene expressions. The management of EPE and ROS metabolism within chloroplast is therefore one of the prime factors for plant survival and performance. Several strategies are adapted by the plant to combat inevitable consequence of oxidative stress induced by EPE and other forms of environmental odds. Photoinhibition, xanthophyll cycle, ascorbate-glutathione pathway, etc. operate efficiently to combat oxidative stress and management of EPE. In all those cases redox signaling pathways are involved in one way or another. Therefore, the elucidation of chloroplastic redox signaling pathway related to maintenance of ROS homeostasis could provide useful information for unfolding the molecular mechanism associated with redox regulation in photosynthetic cell.

References

Alscher RG, Hess JL (1993) Antioxidants in higher plants. CRC Press, Boca Raton

Asada K (1999) The water-water cycle in chloroplasts: scavenging of active oxygen species and dissipation of excess photons. Annu Rev Plant Physiol Plant Mol Biol 50:601–639

Baier M, Noctor G, Foyer CH, Dietz KJ (2000) Antisense suppression of 2-cysteine peroxiredoxin in Arabidopsis specifically enhances the activities and expression of enzymes associated with ascorbate metabolism but not glutathione metabolism. Plant Physiol 124:823–832

Bartoli CG, Simontacchi M, Eduardo T, Beltrano J, Montaldi E, Pantarulo S (1999) Drought and watering-dependent oxidative stress: effect on antioxidant content in *Triticum aestivum* L. leaves. J Exp Bot 50:375–383

Bhattacharjee S (2012) An inductive pulse of hydrogen peroxide pretreatment restores redox homeostasis and mitigates oxidative membrane damage under extremes of temperature in two rice cultivars (*Oryza sativa* L., Cultivars Ratna and SR 26B). Plant Growth Regul 68:395–410

Bhattacharjee S (2014) Membrane lipid peroxidation and its conflict of interest: the two faces of oxidative stress. Curr Sci 107:1811–1823

Bilger W, Björkman O (1994) Relationships among violaxanthin deepoxidation, thylakoid membrane conformation and non-photochemical chlorophyll fluorescence quenching in leaves of cotton (*Gossypium hirsutum* L.). Planta 193:238–246

Bouvier F, d'Harlingue A, Hugueney P, Marin E, Marion-Poll A, Camara B (1996) Xanthophyll biosynthesis: cloning, expression, functional reconstitution and regulation of β-cyclohexenyl carotenoid epoxidase from pepper (*Capsicum annuum*). J Biol Chem 271:28861–28867

Bratt CE, Arvidsson P-O, Carlsson M, Åkerlund H-E (1995) Regulation of violaxanthin de-epoxidase activity by p^H and ascorbate concentration. Photosynth Res 45:169–175

Bugos RC, Yamamoto HY (1996) Molecular cloning of violaxanthin de-epoxidase from romaine lettuce and expression in *Escherichia coli*. Proc Natl Acad Sci U S A 93:6320–6325

Bugos RC, Hieber AD, Yamamoto HY (1998) Xanthophyll-cycle enzymes are members of the lipocalin family, the first identified from plants. J Biol Chem 273:15321–15324

Della Penna D, Pogson BJ (2006) Vitamin synthesis in plants: tocopherols and carotenoids. Annu Rev Plant Biol 57:711–738

Dietz KJ (2011) Peroxiredoxins in plants and cyanobacteria. Antioxid Redox Signal 15:1129–1159

Dietz KJ, Horling F, Konig J, Baier M (2002) The function of the chloroplast 2-cysteine peroxiredoxin in peroxide detoxification and its regulation. J Exp Bot 53:1321–1329

Foyer CH (1996) Oxygen metabolism and electron transport in photosynthesis. In: Scandalios J (ed) Oxidative stress and the molecular biology of antioxidant defenses. Cold Spring Harbor Laboratory Press, New York, pp 587–621

Foyer CH (1997) Oxygen metabolism and electron transport in photosynthesis. In: Scandalios J (ed) The molecular biology of free radical scavenging systems. Cold Spring Harbor Laboratory Press, New York, pp 587–621

Foyer CH (2015) Redox homeostasis: opening up ascorbate transport. Nat Plants 1:14012

Foyer CH, Halliwell B (1976) The presence of glutathione and glutathione reductase in chloroplasts: a proposed role in ascorbic acid metabolism. Planta 133:21–25

Foyer CH, Harbinson J (1999) Relationships between antioxidant metabolism and carotenoids in the regulation of photosynthesis. In: Frank HA, Young AJ, Britton G, Cogdell RJ (eds) The photochemistry of carotenoids. Kluwer Academic Publishers, Dordrecht, p 305

Foyer CH, Noctor G (2003) Redox sensing and signalling associated with reactive oxygen in chloroplasts, peroxisomes and mitochondria. Physiol Plant 119:355–364

Foyer CH, Noctor G (2009) Redox regulation in photosynthetic organisms: signaling, acclimation, and practical implications. Antioxid Redox Signal 11:861–905

Foyer CH, Shigeoka S (2011) Understanding oxidative stress and antioxidant functions to enhance photosynthesis. Plant Physiol 155:93–100

Foyer CH, Lopez-Delgado H, Dat JF, Scott IM (1997) Hydrogen peroxide and glutathione associated mechanisms of acclimatory stress tolerance and signalling. Physiol Planta 100:241–254

Genty B, Harbinson J (1996) Regulation of light utilization for photosynthetic electron transport. In: Baker NR (ed) Photosynthesis and the environment. Kluwer Academic Publishers, Dordrecht, pp 67–99

Gilmore AM (1997) Mechanistic aspects of xanthophyll cycle-dependent photoprotection in higher plant chloroplasts and leaves. Physiol Planta 99:197–209

Hager A, Holocher K (1994) Localization of the xanthophyll-cycle enzyme violaxanthin deepoxidase within the thylakoid lumen and abolition of its mobility by a (light-dependent) p^H decrease. Planta 192:581–589

Havaux M, Bonfils J-P, Lütz C, Niyogi KK (2000) Photodamage of the photosynthetic apparatus and its dependence on the leaf developmental stage in the *npq1* Arabidopsis mutant deficient in the xanthophyll cycle enzyme violaxanthin de-epoxidase. Plant Physiol 124:273–284

Hideg É, Kalai T, Hideg K, Vass I (1998) Photoinhibition of photosynthesis in vivo results in singlet oxygen production detection via nitroxide-induced fluorescence quenching in broad bean leaves. Biochemistry 37:11405–11411

Hideg E, Barta C, Kalai T, Vass M, Hideg K, Asada K (2002) Detection of singlet oxygen and superoxide with fluorescence sensors in leaves under stress by photoinhibition or UV radiation. Plant Cell Physiol 43:1154–1164

Huner NPA, Öquist G, Sarhan F (1998) Energy balance and acclimation to light and cold. Trends Plant Sci 3:224–230

Ishikawa T, Shigeoka S (2008) Recent advances in ascorbate biosynthesis and the physiological significance of ascorbate peroxidase in photosynthesizing organisms. Biosci Biotechnol Biochem 72:1143–1154

Kanwischer M, Porfirova S, Bergmüller E, Dörmann P (2005) Alterations in tocopherol cyclase activity in transgenic and mutant plants of Arabidopsis affect tocopherol content, tocopherol composition, and oxidative stress. Plant Physiol 137:713–723

Krause GH (1988) Photoinhibition of photosynthesis. An evaluation of damaging and protective mechanisms. Physiol Plant 74:566–574

Krause GH (1994) Photoinhibition induced by low temperatures. In: Baker NR, Bowyer JR (eds) Photoinhibition of photosynthesis. Bios Scientific, Oxford, pp 331–342

Krieger-Liszkay A, Fufezan C, Trebst A (2008) Singlet oxygen production in photosystem II and related protection mechanism. Photosynth Res 98:551–564

Mehler AH (1951) Studies of reactions of illuminated chloroplasts. 10. Mechanics of reduction of oxygen and other Hill reagents. Arch Biochem Biophys 33:65–77

Munekage YN, Genty B, Peltier G (2008) Effect of PGR5 impairment on photosynthesis and growth in *Arabidopsis thaliana*. Plant Cell Physiol 49:1688–1698

Niyogi KK, Grossman AR, Björkman O (1998) Arabidopsis mutants define a central role for the xanthophyll cycle in the regulation of photosynthetic energy conversion. Plant Cell 10:1121–1134

Niyogi KK, Shih C, Chow WS, Pogson BJ, DellaPenna D, Björkman O (2001) Photoprotection in a zeaxanthin- and lutein-deficient double mutant of Arabidopsis. Photosyn Res 67:139–145

Noctor G, Foyer CH (1998) Ascorbate and glutathione: keeping active oxygen under control. Annu Rev Plant Physiol 49:249–279

Noctor G, Arisi ACM, Jouanin L, Kunert KJ, Rennenberg H, Foyer CH (1998) Glutathione: biosynthesis, metabolism and relationship to stress tolerance explored in transformed plants. J Exp Bot 49:623–647

Ohad I, Kyle DJ, Arntzen CJ (1984) Membrane protein damage and repair: removal and replacement of inactivated 32-kilodalton polypeptides in chloroplast membranes. J Cell Biol 270:14919–14927

Queval G, Hager J, Gakie're B, Noctor G (2008) Why are literature data for H_2O_2 contents so variable? A discussion of potential difficulties in the quantitative assay of leaf extracts. J Exp Bot 59:135–146

Rumeau D, Peltier G, Cournac L (2007) Chlororespiration and cyclic electron flow around PSI during photosynthesis and plant stress response. Plant Cell Environ 30:1041–1051

Russell AW, Critchley C, Robinson SA, Franklin LA, Seaton GGR, Chow W-S, Anderson J, Osmond CB (1995) Photosystem II regulation and dynamics of the chloroplast D1 protein in *Arabidopsis* leaves during photosynthesis and photoinhibition. Plant Physiol 107:943–952

Santos M, Gousseau H, Lister C, Foyer C, Creissen G, Mullineaux P (1996) Cytosolic ascorbate peroxidase from *Arabidopsis thaliana* L. is encoded by a small multigene family. Planta 1981(1):64–69

Scheibe R, Backhausen JE, Emmerlich V, Holtgrefe S (2005) Strategies to maintain redox homeostasis during photosynthesis under changing conditions. J Exp Bot 56:1481–1489

Smirnoff N, Conklin PL, Loewus FA (2001) Biosynthesis of ascorbic acid in plants: a renaissance. Annu Rev Plant Physiol Plant Mol Biol 52:437–467

Tardy F, Havaux M (1997) Thylakoid membrane fluidity and thermostability during the operation of the xanthophyll cycle in higher-plant chloroplasts. Biochim Biophys Acta 4 1330(2):179–193

Vass I, Aro EM (2008) Photoinhibition of photosynthetic electron transport. In: Renger G (ed) Primary processes of photosynthesis: basic principles and apparatus. Royal Society of Chemistry, Cambridge, pp 393–411

Vass I, Cser K (2009) Janus-faced charge recombinations in photosystem II photoinhibition. Trends Plant Sci 14:200–205

Vranova E, Inze D, Van Breusegem F (2002) Signal transduction during oxidative stress. J Exp Bot 53:1227–1236

ROS: Central Component of Signaling Network in Plant Cell

6

Abstract

Plants often deliberately generate and exploit reactive oxygen species (ROS) or its secondary breakdown products for a number of processes ranging from cell signaling to gene expression. The cellular language associated with ROS signaling network involves a close coordination of four interacting phenomenons, ranging from ROS sensing, signaling, differential expression of redox-sensitive genes, and influencing stress and developmental responses of the plant. The role of ROS as "second messenger" modulating the activities of specific transcription factors or functional proteins is well elucidated. Apart from its bona fide role in the signaling cascades, ROS often complements, synergizes, and antagonizes several growth regulatory circuits through cross talking with other signaling molecules. In this perspective, understanding the mechanism of ROS sensing associated with subsequent signaling cascades in plant cell is not only fundamental but also of immense practical significance, since this knowledge may contribute significantly in agricultural productivity by better management of environmental and oxidative stress. The position of these prooxidants under environmental stress demonstrated that it is essential for the perception and communication of environmental stimuli and associated developmental processes. The retrograde signaling induced by ROS are extremely significant in maintaining cellular redox homeostasis and controlling systemic signaling cascades associated with stress acclamatory responses. Apart from the direct role of ROS in cell signaling, some of the products of oxidative stress, particularly reactive lipid species (RLS), also represent "biological signals," which do not need preceding activation of genes. In the present chapter, an effort has been made to discuss the mechanism of ROS sensing in the elaborate signaling network of plant cell. The present chapter also explores both the mechanisms of signaling cascade of ROS in plant acclamatory defense processes, controlled cell death, and development. The role of redox-sensitive proteins in ROS signaling, its subsequent regulation of Ca^{2+} homeostasis, and MAPK cascades is also discussed. An additional effort has been made to understand the mechanism of H_2O_2-regulated gene expression in plant cell.

S. Bhattacharjee, *Reactive Oxygen Species in Plant Biology*,
https://doi.org/10.1007/978-81-322-3941-3_6

Keywords
ROS · Signal transduction · Redox sensing · Redox-regulated gene expression · Redox-sensitive proteins · Lipid peroxidation products · Oxylipin

6.1 Introduction

A universal response to unfavorable environmental cues is the change in redox homeostasis of cell through accelerated generation and accumulation of reactive oxygen species (ROS) (Bhattacharjee 2010; Miller et al. 2008, 2009; Gill and Tuteja 2010). ROS are in fact the unwelcome by-products of normal metabolic processes of all aerobic organisms and possess strong oxidizing potential leading to oxidation and damage to almost every important classes of biological molecules (Chowdhury et al. 2016; Bhattacharjee 2005; Halliwell 2006). Produced either through incomplete reduction of oxygen ($O_2{\cdot}^-$, $HO_2{\cdot}^-$, H_2O_2, $OH{\cdot}$) or through energy transfer mechanism to triplet ground state of oxygen ($^1O_2$) or even as the secondary oxidative stress products (ROO·, RO·), ROS exhibit their toxic effect particularly under excess titer. When exposed to unfavorable environmental cues, the endogenous levels of ROS can rise significantly, leading to redox imbalance and oxidative stress (Apel and Hirt 2004; You and Chan 2015). To combat oxidative stress, plants have developed and elaborate antioxidative defense system to control cellular titer of ROS. Surprisingly, several informations are poured in the last decade showing evidences where plants have also developed a way to make use of nontoxic lower titer of these prooxidants as signaling component to regulate wide variety of plant processes and their performances under stress (You and Chan 2015; Foreman et al. 2003). This event is basically carried out by redox sensing and signaling pathways through management of spatiotemporal titer of ROS by buffering the two interacting events of ROS biology, i.e., generation of ROS and antioxidant-based detox-scavenging devices that control the endogenous titer of ROS according to the status and need of cellular physiology. In this regard, extensive literature on comprehensive overviews on the positive signaling role of ROS and redox regulation are available (Bhattacharjee 2010; Miller et al. 2008, 2009; Gill and Tuteja 2010; Kovalchuk 2010). Though the concept of redox regulation of plant physiological processes is in vogue for more than a decade, recent time only witnessed direct molecular genetic evidences of the role of ROS in redox sensing, signaling, and gene expression, particularly for the plants undergoing stress acclimation (Miller et al. 2008, 2009; Gill and Tuteja 2010). The most handy example in this regard is the retrograde signaling mediated by auto propagating ROS wave, which is capable of moving at a very fast rate (approximately under abiotic stress for conferring systemic responses necessary for stress acclimation (Miller et al. 2009)).

The tight regulation of the endogenous titer of H_2O_2 is therefore a prerequisite for controlling plethora of cellular processes, from cell damage to protection. Although there are several sources of H_2O_2 in plant cell, but essentially these are all consequences of enhanced rate of metabolism (Fig. 6.1). Out of several sources, the most significant sources of H_2O_2 include photosynthetic carbon oxidation cycle (PCOC) or C2 cycle. Oxidative burst associated with the sudden rise in concentration of H_2O_2 occurs mainly due to active generation of H_2O_2 as a part of HR (hypersensitive response) to infectious organisms. One of the prime sources of ROS, particularly H_2O_2, is NADPH-dependent oxidases or respiratory burst oxidase homologs (Rboh), associated with membrane. This class of proteins is enzymes which are largely regulated by Rho-like proteins (Rho-related GTPases) in plants (Agrawal et al. 2003). Peroxidases are another class of enzymes, linked with the formation of H_2O_2 in cell wall (Bolwell et al. 1995). Surprisingly, when we compare the kinetics of ROS

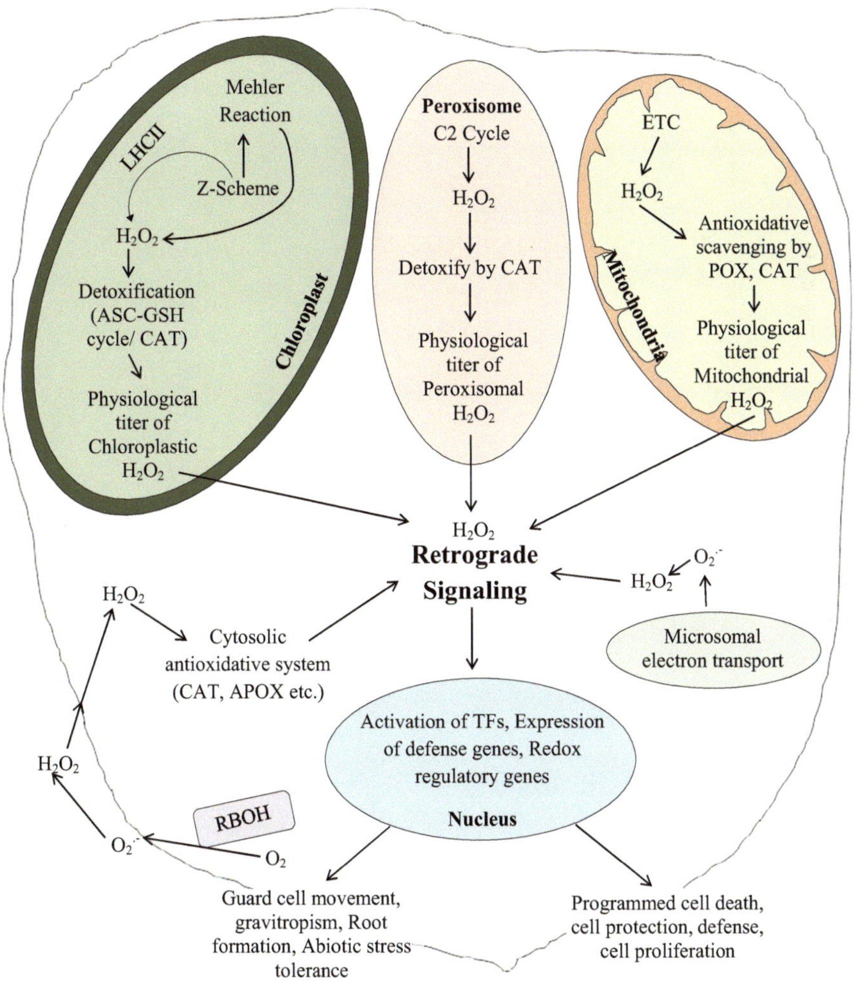

Fig. 6.1 Topology of subcellular generation of H_2O_2 in plant cell showing the organeller contribution and subsequent role of H_2O_2-associated retrograde signaling in the regulation of gene expression in plant cell

formation, it is found that chloroplastic H_2O_2 may be 30–100 times faster than the mitochondrial H_2O_2 (Fig. 6.2) (Robson and Vanlerberghe 2002; Maxwell et al. 1999). Similarly, an efficient array of antioxidant network is constantly on alert for regulating the endogenous titer of H_2O_2 by providing effective scavenging mechanisms (Apel and Hirt 2004; Miller et al. 2010). In fact the turnover of H_2O_2 in plant cells seems to be integrated in a network and is responsible for its "biological effect." The site of production and the developmental state of its genesis are also important for regulating the biological activity of H_2O_2 (Bhattacharjee 2012; Gechev and Hille 2005).

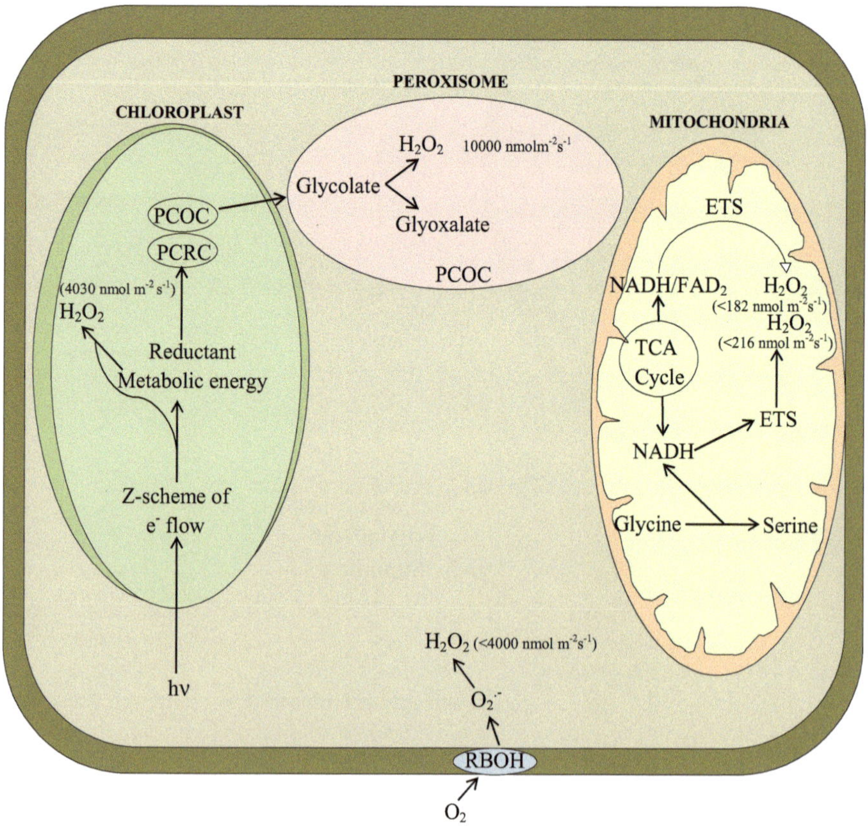

Fig. 6.2 Typical rates of generation of reactive oxygen species H_2O_2 in different compartments of a green cell

6.2 ROS as Signaling Molecule with Specific Advantages

The chemistry and physiology of ROS, H_2O_2 in particular, in plant cell exhibit features typical for second messengers with selective advantages. The following points substantiate the advantages associated with ROS as signaling molecule:

(i) Dynamic changes of the titer of H_2O_2 due to its simultaneous production and scavenging. A simple tilt of balance between H_2O_2 generation and its scavenging by antioxidants (CAT and APOX) changes the physiological titer of H_2O_2 and its subsequent action.
(ii) A significant control over subcellular localization of H_2O_2 signals in cell is possible through organelle-specific generation (chloroplastic, peroxisomal, and mitochondrial), ultimately making spatial control highly specific.
(iii) The most important selective advantages of H_2O_2 signaling depends on their rapid long-distance, self-propagating nature of signal migration. It was recently reported that in *Arabidopsis*, the ROS signal can propagate at a rate of 8.4 cm/min (Mittler et al. 2011).
(iv) The ability of H_2O_2 to interact with hormonal pathways through Ca^{2+} and protein phosphorylation cascades further substantiates the signaling advantages of H_2O_2 (Mittler et al. 2011).
(v) The tight link between H_2O_2 metabolisms, cellular homeostasis, and metabolism is another key signaling advantage associated with H_2O_2 linked redox signaling.

6.2.1 ROS (H_2O_2) Transport

Transportation of ROS, particularly H_2O_2, also strongly supports their signaling role. Moderately high half-life and the rapid diffusion properties across cell membrane enable H_2O_2 molecule to facilitate its signaling function. H_2O_2 generated in chloroplast may be spilled out of chloroplast in a light-dependent mechanism (Mubarakshina et al. 2010). In fact H_2O_2 is passaged cell organelles like chloroplast, mitochondria, peroxisome through specific homologs of aquaporin plasma membrane intrinsic proteins, or tonoplast intrinsic proteins (TIP and PIP family trafficking proteins). Expression analysis of *Arabidopsis thaliana* genes, like AtTIP1;1 AtTIP2;2 in yeast reveals that, H_2O_2 causes significant reduction in survival rate (Bienert et al. 2007). Moreover, specific conserved sites in PIP2 are found to be involved for H_2O_2 transportation across the membrane (Dynowski et al. 2008). H_2O_2 was also found to have significant impact on general osmotic phenomena in a concentration-dependent manner (Benabdellah et al. 2009).

6.3 ROS Signal with Specific Targets in Plant Cell

Redox signals at metabolic interface are the most fundamental forms of information at cellular level and are monitored by the plants involving specific signaling cascades (Mylona and Polidoros 2010; Mahalingam and Fedoroff 2003). More complex aspects of redox regulation of physiology of plants eventually through the regulation of gene expression developed with the gradual evolution of higher plants. It is now generally accepted that redox signals are the key regulators of plant metabolism, growth, and development and may have even ample of cross talking with other systems of signal transduction of plants. In fact, controlled production and maintenance of titer of H_2O_2 acts as "second messenger" along with other mediators like Ca^{2+}, not only in plant responses to environmental stress but also in hormone signaling (Bhattacharjee 2005; Locato et al. 2010).

In order to have a comprehensive idea regarding the H_2O_2-mediated signal transduction, one has to develop the principal understanding of the fact that how the variable titer of H_2O_2 is sensed. One realistic option is the direct oxidative modification of transcription factor with redox-sensitive groups (Allen and Tresini 2002). But, in actuality, there seems to be a great deal of additional complex redox sensing in different signal transduction routes. In fact, the H_2O_2-associated redox changes in the diverse organelles (mainly chloroplast and mitochondria) are signaled to the nucleus. Surprisingly at the onset, the "sensing" function is generally performed by the antioxidative system itself. This system also acts as a strong buffer against any effort that changes the titer of H_2O_2, maintaining relatively low concentrations of oxidants under normal optimal conditions. The redox homeostasis of the cell is generally perturbed without large changes in the concentration of H_2O_2. This is feasible because of the presence of plethora of components capable of scavenging H_2O_2. In case of catalase, mutants placed under photooxidative stress, or excess photochemical energy, H_2O_2 concentration is not greatly increased, relative to the wild type, but the glutathione pool is found to be massively perturbed (Allen and Tresini 2002; Neill et al. 1999). The enhanced availability of ROS may therefore be sensed by the cell as increased oxidative flux through key components, rather than marked increase of H_2O_2 titer. This view strongly supports the existence of a dynamic system for allowing acclamatory changes through the components of antioxidative system that are integrated into the signal transduction network. It would allow appropriate response to occur as a result of increased flux of H_2O_2, even in absence of marked changes in H_2O_2 titer. Apart from that, changes in ROS trigger marked modulation in the expression of gene far beyond their perception by antioxidative systems (Mylona and Polidoros 2010; Mahalingam and Fedoroff 2003). This ultimately bestowed upregulation of defense system against environmental stresses. Therefore, antioxidative compounds have dual role of scavenging and signaling (through sensor scavenging). In these cases, the antioxidants probably exhibit low capacity of antioxidative role compared to classical detoxification scavenging.

Works related to the chemical nature of "sensor-scavenging component" (Fig. 6.3) revealed that heme-based enzymatic antioxidants (catalase, ascorbate peroxidases) and thiol-containing antioxidant (GSH) are basically the candidates for

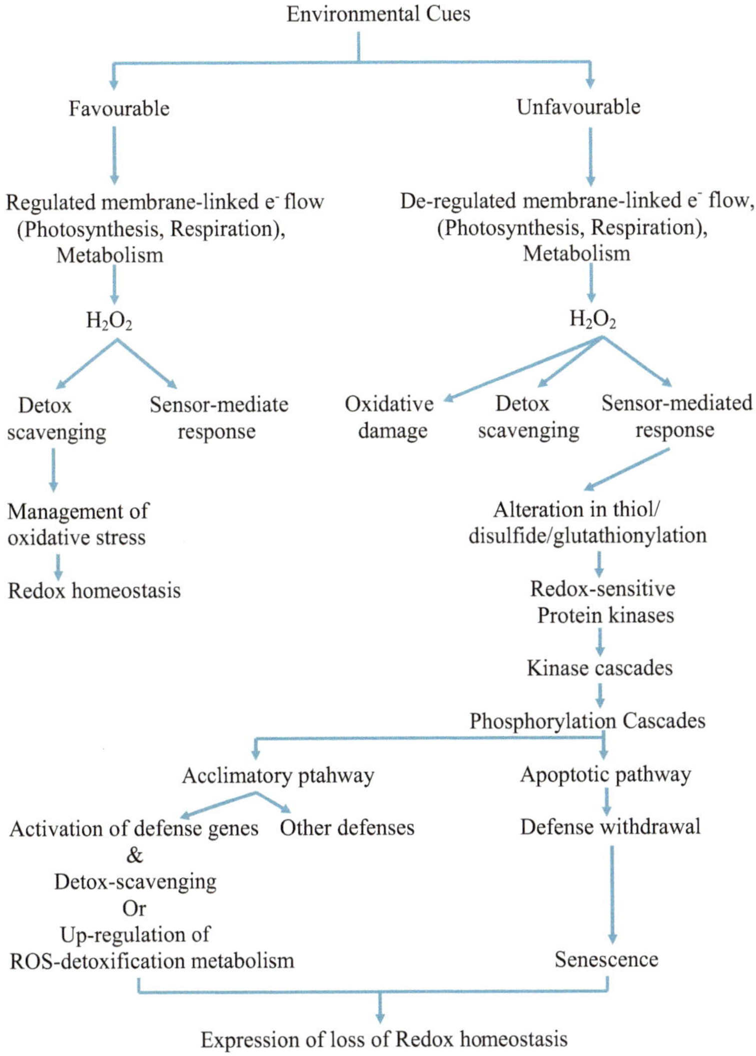

Fig. 6.3 Figure explaining the perception of normal and enhanced titer of H_2O_2 involving redox components or antioxidant system. Under normal growing condition, H_2O_2 are produced as a consequence of many metabolic events and are efficiently removed by detox scavenging of antioxidants. Under unfavorable environmental cues, increased production of H_2O_2 or oxidative burst causes increased oxidation of "sensor-scavenging" redox components locked into signal transduction pathways. Signaling pathway operates via kinase cascades and other secondary messengers that ultimately lead to upregulation of ROS detoxification capacity (acclimatory pathway) or cell death pathway involving defense withdrawal. The decisive factors that determine the feasibility of acclimatory pathway or cell death pathway are yet to be explored that could determine the intensity of H_2O_2 signal and their location

such responses. Plants also contain several proteins with redox-active thiol groups, some of which have been shown to have activity against peroxides. These include chloroplastic and cytosolic glutathione peroxidases, chloroplastic and cytosolic peroxiredoxins, glutaredoxins, and chloroplastic thioredoxins (Noctor and Foyer 1999; Rouhier et al. 2009; Nishiyama et al. 2001). Although the exact roles of these thiol-containing redox sensors or ROS scavengers remain to be elucidated, it is clear that there may be considerable discrepancies of function within each class of these molecules.

6.3.1 Redox Sensors of Plant Cell

Changes in environmental cues and infection, in general, cause transient rise in H_2O_2 and hence can be viewed as a signal that relays the messages associated with changes in external environmental factor to downstream effectors. The existence of H_2O_2 sensors still remains elusive, although it is presumed that enhanced H_2O_2 levels could be perceived directly by redox-sensitive proteins or transcription factors that orchestrate downstream signaling cascades and amplification of response (Miller et al. 2008; Veal et al. 2007; Marinho et al. 2014). One of the important prerequisites for H_2O_2 sensing and subsequent signaling is the availability of physiological titer of H_2O_2 by escaping its decomposition by antioxidants. Broadly speaking, there exist several molecular mechanisms of H_2O_2 sensing in plant cell (Fig. 6.4). The basic device of H_2O_2 sensing in plant cell involves thiol and non-thiol protein-based mechanisms, apart from antioxidant-coupled signaling (Veal et al. 2007; Marinho et al. 2014; Kim et al. 2002; Lee et al. 2004; Tonks 2005).

To characterize the bona fide sensors for H_2O_2, one has to apply the basic norms of chemistry. It convey the fact that primarily the low molecular weight thiols are preferred candidates for H_2O_2 sensing and the kinetics of reaction of low molecular weight thiols with H_2O_2, as exemplified by rate constants for H_2O_2 with cysteine and GSH (2.9 $M^{-1}S^{-1}$ and 0.87 $M^{-1}S^{-1}$, respectively), support their slow but steady reaction (Ferrer-Sueta et al. 2011). In proteins the electrostatic environment around the free-SH groups, particularly of cysteine, renders these groups more acidic, thereby exhibiting more reactivity with H_2O_2. In several proteins, reduced stabilization of thiolate in cysteine residues enhances the nucleophilicity of thiolate by several order of magnitude rate constants of H_2O_2 (Ferrer-Sueta et al. 2011). Therefore, the very concept of H_2O_2 sensing and subsequent signaling was proposed following the discovery of cysteine containing proteins as signaling intermediates like phosphatases, kinases, transcription factors, etc. In fact, the –SH groups of those molecules cause changes in biological activity. Therefore, according to this paradigm, with any changes in physiological titer of H_2O_2, these proteins became the target of oxidation, and subsequent cascades of molecular events commences. One of the bottlenecks of identifying the sensors, particularly in plant cell, is lack of data related to quantitative kinetics characterizing the reactivity of H_2O_2 with target proteins, in spite of wealth of data identifying reactive SH groups (Marinho et al. 2014; Ferrer-Sueta et al. 2011). Therefore, based on the available information on H_2O_2 reactivity

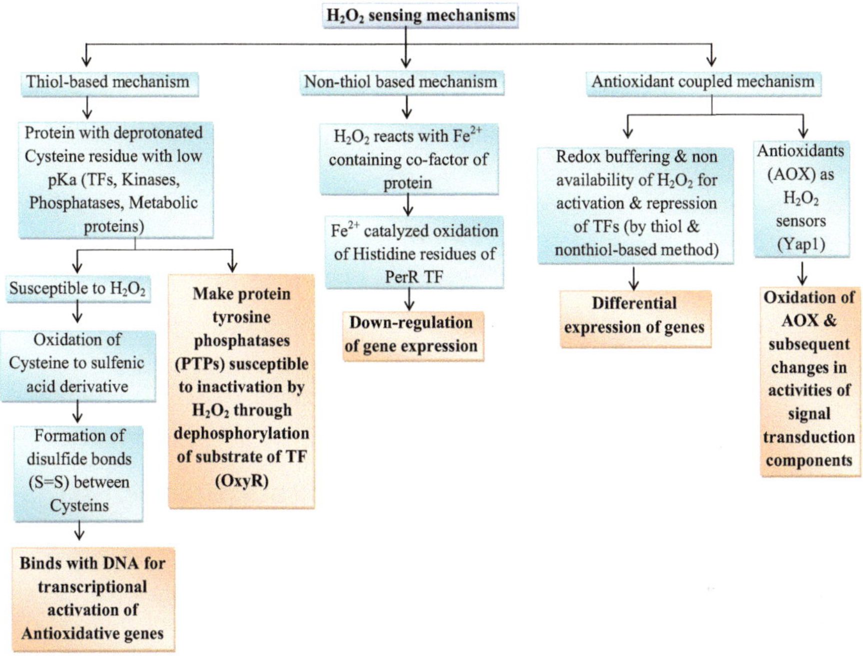

Fig. 6.4 Different modes of H_2O_2-sensing mechanisms [thiol, non-thiol, and antioxidant coupled] in plant cell. (Detail in text)

with target proteins and the subsequent signaling, a sensor may be characterized by the following features:

1. A sensor should be sensitive to changes in endogenous titer of H_2O_2 or oxidant. For example, in case of protein thiols, the reaction between H_2O_2 and thiol compounds is a second-order reaction, vouching the fact that reaction depends on H_2O_2.
2. A sensor should have dynamic characteristics that suit its function, implying the fact that before the termination of H_2O_2 signal, the sensor should be activated once again to start another round of activation by H_2O_2.
3. A sensor may function along with performing other tasks, as in the case of antioxidant-coupled sensing.
4. Thiol-based protein sensors should remain in deprotonated state.

One of the basic prerequisite of H_2O_2-sensing mechanism is the maintenance of specific titer of H_2O_2 by evading or controlling its decomposition through antioxidative defense. In thiol-based redox sensing mechanism, thiol-containing proteins render them susceptible to oxidation by H_2O_2. The low pH of the cytosol causes protonation of cysteine residues of proteins, making them insensitive to H_2O_2. So, the utilization of H_2O_2 in signal transduction event involving thiol compounds (cysteine compounds with low pKa) as sensors largely depends on cytosolic pH. The thiol protein sensors that are susceptible to oxidation by H_2O_2 behave in such a way that the deprotonated cysteine residues can detect H_2O_2. In fact, the exposure to H_2O_2 and other oxidants in cell leads to reversible oxidation of thiol groups of the key cysteine residues of several signaling intermediates including kinases, phosphatases, transcription factors, metabolic enzymes, ion channels, antioxidants, etc., eventually affecting the signaling procedure.

From different works it became clear that H_2O_2 signaling can operate in two ways (Fig. 6.4):

(i) A transient localized higher titer of H_2O_2-specific signaling that activate sensors with low reactivity and inhibition of peroxiredoxin activity
(ii) H_2O_2 signaling mediated by initial sensor proteins with high reactivity toward H_2O_2 that subsequently activate a low-reactive protein of signaling pathway

6.3.2 Some Recognized H_2O_2 Sensors of Plant Cell (OxyR, Hsf1, PerR, PTPs, HSFs, YAP-1)

One of the best characterized transcriptional activator characterized as H_2O_2 sensor which get directly oxidized in response to elevated concentration of H_2O_2 is OxyR (a member of LysR family). Though both the reduced and oxidized form of OxyR is capable of binding with DNA, only the oxidized form of OxyR is capable of triggering transcriptional regulation of antioxidant genes (Marinho et al. 2014; Storz et al. 1990). Basically the TF contains the DNA-binding central recognition or response domain and the C-terminal domain required for activation (Schell 1993). In oxidized state OxyR contacts the DNA motif in major groves on one face of DNA helix, whereas in reduced form, it binds DNA in two pairs of major grooves separated by helical turn (Storz et al. 1990); the most critical event in sensing H_2O_2 by OxyR is the oxidation of cysteine 199 to sulfenic acid derivative followed by the formation of disulfide bond between cysteine 199 and 208 (Kim et al. 2002; Storz et al. 1990). The reduction and downregulation of OxyR also causes a negative feedback loop involving the OxyR-regulated expression of gene encoding Glutaredoxin-1 (Zheng et al. 1998).

The homotrimeric DNA-binding TF, heat shock factor 1 (HSF1), a member of HS transit factor family that binds with heat shock element (HSE), is found to be another redox-sensitive TF. In most of the cases, HSEs contain two more inverted repeat sequences of 5-bp nGAAn. Apart from its involvement in heat shock responses, Hsf1 is also found to be associated in redox response (Marinho et al. 2014; Yamamoto et al. 2007) through upregulation of genes for molecular chaperones, important defense enzymes, etc. (Marinho et al. 2014; Yamamoto et al. 2005).

PerR (peroxide regulon repressor) – a metal-dependant TF – is a major redox sensor induced by H_2O_2 (Helmann et al. 2003). Though discovered as ferric uptake repressor (Fur) proteins, its role has been proved to be basically as H_2O_2 sensor and oxidative stress response (Marinho et al. 2014). In several cases PerR is found to be a functional equivalent for OxyR and substitute PerR in gram-positive bacteria

(Eggler et al. 2005) in redox sensing mechanisms (Dubbs and Mongkolsuk 2012). PerR DNA binding is primarily activated by metal cations and interacts with a specific palindromic sequence, residing close to promoter sequence of target genes. The upregulation of several redox-sensitive genes have been noticed under the activation of PerR, influencing oxidative stress response (Helmann et al. 2003).

Another important H_2O_2 sensor, protein tyrosine phosphatases (PTPs), that plays crucial role in signal transduction event under unfavorable environmental conditions catalyzes dephosphorylation of tyrosine residue, causing inactivation of MAPKs. The deprotonated cysteine residues in the catalytic dephosphorylation of substrate render PTPs susceptible to inactivation by H_2O_2. Several evidences showed the role of PTPs under stress through its oxidation and subsequent inhibition by the second messenger H_2O_2. The nature of oxidation of PTPs and the products form thereafter varies. The reversible sequestration of the deprotonated cysteine residue into an oxidized form regulates different families of phosphatases (vanMontfort et al. 2003).

Although H_2O_2-sensing mechanisms basically operates involving thiol compounds with reversible oxidation of deprotonated cysteine residues, but in some cases, H_2O_2 directly reacts with Fe^{2+} containing cofactors of some TFs or proteins. In these non-thiol mechanisms, involving metal-catalyzed oxidation of histidine residues is important for H_2O_2 sensing by TFs like PerR (Helmann et al. 2003). PerR in general contains a structural Zn^{2+}-binding and regulatory Fe^{2+}-binding sites (Dubbs and Mongkolsuk 2012). The Fe-binding site binds DNA, repressing the expression of target genes (Dubbs and Mongkolsuk 2012).

Heat shock factors (HSFs) which are shown to be responsive to oxidative stress are the most important candidate for H_2O_2 sensors in plant (Marinho et al. 2014). In animals, it was found that HSFs are influenced by H_2O_2 titer and subsequently form DNA-binding competent homotrimers (Ahn and Thiele 2003). Although not completely elucidated, plants may utilize the similar pathway of H_2O_2 signaling cascades using HSFs as sensor molecule. Recently, H_2O_2 perception has been explained in terms of H_2O_2-mediated oxidation of methionine (Met) to methionine sulfoxide (MetSo). The oxidation of Met to MetSo provides oxidative signals required for changing protein phosphorylation state (vanMontfort et al. 2003). The in vitro oxidation of Met^{538} in leaf glutathione reductase strongly inhibits phosphorylation of adjacent ser^{534}. Application of H_2O_2, on the other hand, causes reduction of Ser^{534} phosphorylation, while the upregulation in expression of cytosolic MetSo repair enzyme has the contrasting effect. So, the experimental evidence strongly supports the existence of redox-methionine-dependent switch for integrating H_2O_2-mediated oxidative signal in plant cell. Recently, several proteomic approaches (2D-LCMS, 2D-MALDI TOF, COFRADIC) were used successfully for identifying oxidized methionine residues (Ghesquière et al. 2011).

YAP1 is a basic zipper type transcription factor which induces several redox-sensitive genes under oxidative stress. ROS like H_2O_2 augment the accumulation of nuclear YAP-1 by oxidizing cysteine residue from intramolecular disulfide bond that appears to trap YAP-1 in the nucleus, thereby enhancing their reactivity (Marinho et al. 2014). *Arabidopsis* genome does not appear to have sequence similar to YAP-1, but functionally comparable elements do present in the early perception of changes in redox state of plant. Yeast AP1 (YAP1) is in fact one of the members of yeast activator protein (yap) family that comprises eight members (Fernandes et al. 1997). All members display a significant sequence similarity at DNA-binding domains.

Peroxiredoxins (PRX), otherwise recognized as thioredoxin (THX) peroxides, have been revealed to reduce peroxides using the reducing power of thioredoxins (Trx). The enzyme system is intimately connected to Z-scheme of photosynthesis via a specific chloroplastic Trx isoform. In fact, the primary role of these thiol-containing redox-sensitive proteins is not to detoxify or scavenge peroxides but to sense enhanced formation or in other words to kick off a signaling event by acting as a "signal-scavenging component" (Fig. 6.3).

An additional device of redox-signal perception involves oxidation of glutathione pool along with the increase in total glutathione pool under environmental stresses (Rouhier et al. 2009). The means of sensing this redox perturbation of plant cell may also involves protein glutathionylation, in which GSH forms a mixed disulfide with target protein, thereby modifying the activities and performances of enzymes and transcription factors. This process is considered to participate in redox signaling and protection of protein structure and function.

One more noteworthy class of thiol-containing protein is certain subclasses of GST superfamily which is dynamic in reducing peroxides or dehydroascorbates (DHA) or in catalyzing thiol transferase reaction (Daniel 1993). Some of these GSTs are significantly upregulated by oxidative stress especially under biotic stresses. So, it seems that the influence of H_2O_2 can be efficiently harnessed to convey necessary redox information. These signals in most of the cases are incorporated into complex redox network that involves electron carriers (plastoquinone, ubiquinone, plastocyanin) and electron acceptors (ferredoxin, thioredoxin, glutaredoxin).

6.3.3 Role of Redox-Sensitive Proteins (RSPs)

Plant cells have unique capacity to sense and transduce the ROS, particularly the H_2O_2 signals into responses, which are largely mediated by redox-sensitive proteins or sensors. This class of proteins mainly functions through switching "on" and "off" the function of reversible oxidation/reduction cascades, depending on the cellular redox state. ROS H_2O_2 can get the capacity to oxidize the redox-sensitive proteins directly or indirectly via some other important low molecular weight redox-sensitive molecules like glutathione, glutaredoxins (GRXs), peroxiredoxins (PRXs), or thioredoxins (THX) which largely control the internal redox cues of the cell (Mylona and Polidoros 2010; Rouhier et al. 2009). So, RSPs are susceptible to oxidation and reduction that depend on the endogenous concentration of H_2O_2 and other ROS and/or redox state of the cell. Further these RSPs accomplish their signaling role via other downstream components like phosphatases, kinases, and even the transcription factors (TFs). In some cases ROS directly oxidize PRX and TRX and eventually modulate the activities of TFs (Mylona and Polidoros 2010; Nishiyama et al. 2001). In fact, it is also found that the significant level of redox regulation of gene expression is controlled by TRXs, GRXs, PRXs, and protein disulfide isomerases (Rouhier et al. 2009). TRXs are small (approximately12 kDa) protein with S=S reducing activity. They oxidized directly by ROS or indirectly by PRXs. Thioredoxin reductase can directly reduce TRXs by activating NADPH-dependent enzymes. There are ample evidences to suggest that TRXS and other similar redox-sensitive proteins are enzymatic mediators of redox regulation at transcriptional levels (Mahalingam and Fedoroff 2003). Some study revealed that UV irradiation sponsors translocation of TRX to the nucleus of mammalian cell where it upregulates TFs NF-kB and AP-1 by increasing the DNA-binding property (Mahalingam and Fedoroff 2003). TRX can bind in a straight way with NF-kB p^{50} and indirectly with AP-1 involving other important redox factor1 (Karimpour et al. 2002). In contrary, plants exhibited important elaborate redox-regulating system for controlling expression of redox-sensitive genes (Mahalingam and Fedoroff 2003; Rouhier et al. 2009). The photoreduced ferredoxin (FdH_2) of Z-scheme of electron flow transform the reducing power to TRX by Fd-Trx reductase, which eventually interacts with target enzymes for their activation. The "thioredoxin targets" including the "transcription factors" are extremely sensitive RSPs or RNA-binding proteins which regulate translation or stability of chloroplastic mRNA, particularly under the exposure of excess solar energy involving FRX-TRX state (Mahalingam and Fedoroff 2003).

6.3.4 Redox Signaling and Regulation of Ca^{2+} Homeostasis and MAPK Cascades

Control and regulation of endogenous titer of Ca^{2+} is one of the major targets of ROS signaling pathway (Petrov and Van Breusegem 2012). Several past works support the view that H_2O_2 signaling and Ca^{2+} homeostasis in plant cells are interdependent, exhibiting feed-forward or backward interaction between H_2O_2 metabolism and Ca^2 level. Regulation of stomatal conductance by controlling the turgor pressure of guard cell is probably one of the extensively studied processes to corroborate ROS (H_2O_2)-mediated Ca^{2+} signaling. In fact, dehydration stress-induced ABA-mediated formation of H_2O_2 in stomatal guard cells result in increase in cytosolic Ca^{2+} ($[Ca^{2+}]_{cyt}$) (Pei et al. 2000). The cytosolic Ca^{2+} found to increase in a biphasic manner in oxidative stressed *Arabidopsis* seedlings (Rentel and Knight 2004). The instant Ca^{2+} peak under the inductive pulse of oxidative stress is considered as one of the initial responses of H_2O_2 signaling. The spermidine oxidase-mediated oxidative stress regulates pollen membrane hyperpolarization of pollen membrane by activating Ca^{2+} channels to ensure pollen tube growth (Wu et al. 2010).

On the contrary, the cytoplasmic Ca^{2+} is also able to trigger changes in H_2O_2 level. A constant cytoplasmic influx of Ca^{2+} is required for upregulating the activities of membrane-bound NADPH oxidases (Lamb and Dixon 1997). Infected *Arabidopsis* with blocked NADPH oxidase does not change Ca^{2+} level. Further, the application of a calcium chelator or channel blocker suppresses H_2O_2 accumulation necessary for subsequent hypersensitive response (Grant et al. 2000). These experiments strongly favors the view that Ca^{2+} works upstream of H_2O_2 in signaling cascades (Lachaud et al. 2011).

Ca^{2+} homeostasis also found to control antioxidative defense in plant. Experimental evidences suggest that a rise in the titer of cytoplasmic Ca^{2+} cause significant detoxification of ROS H_2O_2 involving antioxidant enzymes. Exogenous application of Ca^{2+} channel blocker, chelator, and calmodulin inhibitor to germinating *Amaranthus* seeds causes significant downregulation of antioxidative defense enzymes (Bhattacharjee 2008), strongly substantiating Ca^{2+} regulation of oxidative stress in plant tissue. Ca-CaM signaling pathway also regulates a number of different target proteins in signaling cascades including MAP kinase (MAP kinase 8, which in turn, downregulates ROS formation by activating NADP oxidases).

So, as indicated above, ROS signaling seems to be tightly integrated with Ca^{2+} signaling and other signaling networks in plant cell, where both H_2O_2 and ROS accumulation were found to precede the commencement of signaling through these networks or, in some other cases, ROS accretion was found to be an outcome of signaling through these elaborate signaling networks.

MAPK cascades are a good example of H_2O_2-activated signaling network (Fig. 6.5). It is found that several MAP kinase cascades can be activated following H_2O_2 and ROS accumulation. The redox-sensitive MAP kinases, which include MAPKKK, MAPK1, MAPK4, and MAPK6, usually remain extremely active under environmental stress-induced oxidative stress. The active MEKK1 pathway under oxidative stress seems to be the activator of two MAPKKs (MKK1 and MKK2), which in turn activates MAP kinase cascades (Mittler et al. 2011). Mekk1 was suggested to be specifically needed for the operation of MAPk4 by H_2O_2 (Mittler et al. 2011). Another two H_2O_2- and ROS-responsive MAP kinases (MAPkK and MAPK6) which depend on MKK9 for their activity are also reported in plant cell. The expression of MAPK9 and MAPK 12 which got upregulated during ABA-induced stomatal closure generally would function downstream to H_2O_2 signaling. Similarly, MAPK12 was found to be upregulated in response to ABA and H_2O_2 application (Mittler et al. 2011).

6.4 H_2O_2-Responsive Transcription Factors

One of the most important components in the chain of H_2O_2 signaling pathway or associated redox circuit is a variety of redox-sensitive transcription factors. They in general are capable of inducing a massive reprogramming of the transcriptome and lead to the generation of a second wave of effectors in the downstream, which in turn switch on various responses like defensive mechanisms associated with stress acclimation or induction of cell death machinery. The noteworthy TFs that are associated with H_2O_2 signaling are members of diverse families, including WRKY, NAC, ZAT, DREB, bZIP, and MYB (Fig. 6.5).

Balazadeh et al. (2010a) reported that in *Arabidopsis* 15, senescence-associated NAC TFs are upregulated by H_2O_2 during leaf senescence, of which *ANAC032* and

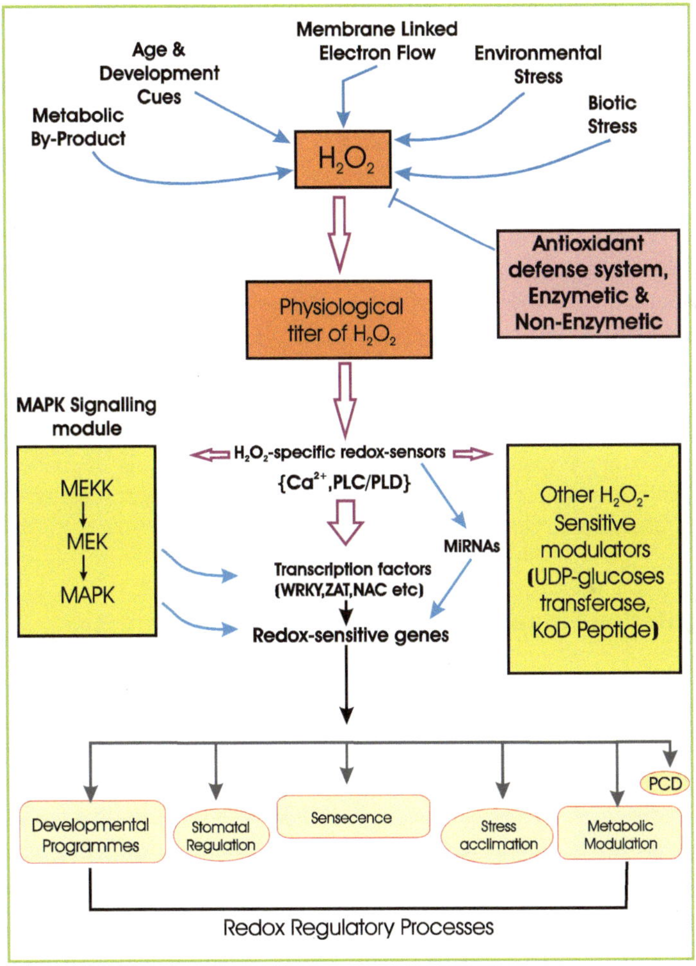

Fig. 6.5 Schematic pathways showing the role of ROS in gene expression (*ROS* reactive oxygen species, *MAPK* mitogen-activated protein kinase, *MAPKK* MAPK kinase, *MAPKKK* MAPKK kinase, *TF* transcription factor, *COTF* transcription cofactor)

ANAC042 are induced around 25 times. The TF ANAC042 has already been reported before as a possible target of H_2O_2 signaling pathways (Gechev et al. 2005), and it was proposed that its activation occurs through the OXI1/MPK3/6 cascade. Another key redox regulator of natural and salinity-induced senescence is the TF ANAC092. H_2O_2, a component of oxidative burst, noticeably plays a significant role in ANAC092-mediated responses (Balazadeh et al. 2010a, b). A similar redox-regulatory network, which includes cross talk between salt and ROS (H_2O_2), was also ascribed to the TF ORS1, the paralogue of ANAC092 (Balazadeh et al. 2011).

TFs WRKY are often described as fundamental players in plant responses to oxidative stresses (Eulgem and Somssich 2007). Of these class of TFs, WRKY52 and WRKY70 are the most extensively studied ROS-inducible representatives of the family in plants. WRKY52, which is a senescence-related transcription factor, can also potentiate the expression of a variety of stress, defense, and senescence-associated genes, when overexpressed, leading to accelerated senescence (Miao et al. 2004). On the other hand, TF WRKY70 is specific for abiotic stress (salinity, dehydration) and shows a similar expression pattern to *Zat7* in *apx1* mutant plants (Miller et al. 2010). It was formerly established that WRKY70, together with the HASTY TF, and Zat7 interact in a yeast two-hybrid assay through the EAR-domain of Zat7 (Ciftci-Yilmaz et al. 2007). Other H_2O_2-induced TFs that are upregulated by environmental stress-induced oxidative stress from different families include DREB2A (Rizhsky et al. 2003).

The zinc-finger TFs from the Zat family are also often related to oxidative burst and associated oxidative stress and H_2O_2 signaling. For example, *Zat11* and *Zat12* are highly induced or upregulated by ROS (Gadjev et al. 2006). In addition, it is also found that overexpression of *Zat12* leads to increased levels of oxidative stress-responsive transcripts, while its reversal results in higher sensitivity to H_2O_2 (Rizhsky et al. 2003; Davletova et al. 2005).

6.5 Redox Signaling, miRNAs, and Gene Expression

The first genome-wide study of H_2O_2-regulated miRNAs in plants discovered seven miRNA families which were found to be differentially expressed (both upregulated and downregulated) after ROS (H_2O_2) treatment in rice seedlings (Li et al. 2011). These differentially expressed miRNAs include miR169, miR397, miR528, miR827, miR1425, miR319a.2, and miR408-5p. Out of these, five are upregulated by H_2O_2 (miR169, miR397, miR827, miR1425, and miR408-5p), and two are downregulated (miR528 and miR319a.2). miRNAs can influence signaling intermediates of growth factors. One of the examples of direct targets of these miRNAs is IAR1 proteins (IAA-alanine resistance protein 1), which is known to be important for auxin homeostasis. Similarly, the miR528 targets HAP2-like TF; miR397 targets laccases, involved in lignin biosynthesis; miR408-5p and miR827 influence major facilitator super-family proteins (a group of secondary transporters).

Characterization of changes in gene expression regulated by loss of redox homeostasis of the cell is of considerable importance for developing stress-tolerant plants. The information regarding regulatory role of ROS on gene expression is primarily the outcome of experiments using mutant and transgenic plants with direct application of ROS or ROS generating or modulating chemicals. Results of these experiments contributed information on ROS-responsive inducible genes. In recent times, a number of genes involved in redox signal transduction, stress response and tolerance, metabolism, and developmental cascades have been identified, ultimately revealing a highly dynamic and redundant network of redox-sensitive genes (Allen 1993, Mylona and Polidoros 2010). In bacteria at least 80 genes are found to be under the influence of redox regulation (Demple 1991). In yeast approximately 300 genes (Godon et al. 1998) and in plants more than 100 genes are found to be differentially regulated by ROS (Stephen et al. 1995), and these numbers are growing steadily with the application of cDNA, microarray technique with selective transcriptomic analysis of oxidative stress-regulated gene expression (Varnova et al. 2002).

One of the examples of induction of defense genes, controlled by redox status or signaling of the plant cell, is glutathione S-transferase (GSTs), which catalyzes the conjugation of GSH to a variety of hydrophobic electrophilic compounds. Compounds with bound GSH undergo cellular detoxification pathway (Daniel 1993). The promoter element consists of two adjacent AP1-like sites (Friling et al. 1992). The consensus sequence of this site (ΔGACA (A/T) (A/T) GC) is called electrophile-responsive element or antioxidant-responsive element. Two adjacent AP1-like sites are also present in the *Arabidopsis* GST 6 gene and constitute promoter element (Chen et al. 1996). This promoter element is at least in part required for GST 6 inductions by H_2O_2, SA, and auxins (Chen et al. 1996). A single antioxidant-responsive element has recently been identified in the promoter of a maize catalase gene (Cat 1) and was found to bind molecular factors from senescing scutella that accumulate Cat 1 transcripts, probably as a result of oxidative stress (Polidoros and Scandalios 1999).

Another example of redox-regulated transcriptional activation is the G box (CACGTG), an omnipresent cis-element present in many plant genes which is thought to mediate response to diverse environmental stimuli involving primarily redox changes (Dröge-Laser et al. 1997). Together with the H box (CCTACC), the G box functions in the activation of phenylpropanoid biosynthetic genes. Transcription of at least two of these genes that encode phenylalanine ammonia-lyase (PAL) and chalcone synthase (CHS) is under redox control (mainly controlled by GSH) (Wingate et al. 1988). Heat shock elements and heat shock factors are also found to participate in redox-regulated gene expression in plants. Activation of heat shock factor is characterized by the conversion of this factor from monomer to trimeric state, a process induced by heat shock and large variety of conditions that generate abnormally folded proteins. Disulfide-linked aggregates of cellular proteins are formed as a consequence of changed intracellular redox homeostasis and are one of the signals required for HSF trimerization (Arrigo 1999). Therefore, two major steps in transcriptional activation of eukaryotic transcription factors seem to be influenced by redox balances: nuclear relocalization and DNA binding.

6.6 ROS Cross Talk with Other Signaling Pathway Under Environmental Stress

To facilitate or exert such an elaborate and precise effect on plant metabolism and cellular physiology, the redox signal with ROS as signaling compound must cross talk with other signaling molecules (Mittler et al. 2011, Bhattacharjee 2012). In other words, redox signaling is highly integrated with other signaling pathways (mainly hormonal) that help the system in processing and transmitting environmental inputs to evoke appropriate molecular and physiological responses, particularly under

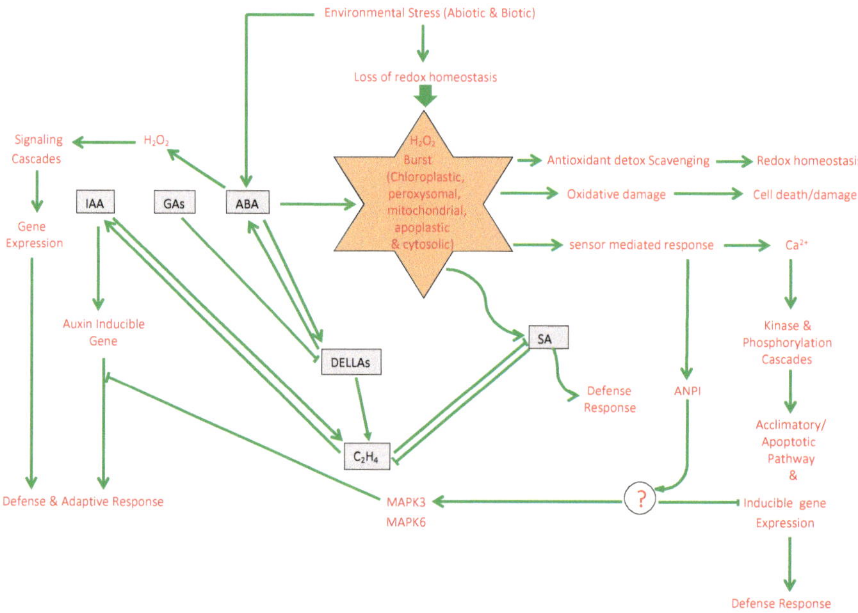

Fig. 6.6 Figure showing the cross talk between ROS and other signaling network (Detail in the text)

environmental odds. Participation of almost all recognized plant growth regulators like IAA, GAs, ABA, SA, and C_2H_4 together with redox signaling (both feed forward and backward) allows the plant not only to regulate short-term acclimation and long-term adaptive responses but the developmental processes as well (Fig. 6.6). The stress-tolerant signaling events regulated by plant hormones which may lead to activation of acclimation responses, such as reduction of stomatal conductance and regulation of hydraulic conductivity, found to be largely mediated by ROS (Miller et al. 2010, Sakamoto 2008). There exists significant cross talk between ROS and hormones like GAs, ABA, IAA, SA, and C_2H_4, which ultimately confer stress acclimation, defense, and responses to cell death and developmental cues (Miller et al. 2010, 2011).

Plant hormones are not only downstream of ROS signal; ROS themselves are secondary messengers in many hormone-mediated signaling pathways (Chen et al. 1993a, b). For example, H_2O_2 was found to be involved in the upregulation of the synthesis of stress hormones such as C_2H_4 and salicylic acid (Leon et al. 1995). Therefore it is conceivable that feedback/feed-forward interactions between ROS and hormones occur (Fig. 6.6).

Natural growth retardant ABA is concerned with a number of abiotic stress responses associated with dehydration, cold, salinity, heat shock, etc. One of the physiological consequences of ABA actions is stomatal closure that downregulates transpiration and water-use efficiency of plants. As a result, availability of CO_2 for fixation of the PCRC (C_3 cycle) cycle is limited, which may, in turn, augment ROS formation in the chloroplast, through Mehler reaction. Previous studies have shown that ABA can change activities of several important antioxidant enzymes and can even affect expression of genes encoding them (Gong et al. 1998; Kurepa et al. 1997). Both H_2O_2 and ABA independently acclimatize maize seedlings to chilling stress; however, the relative position of these two signaling compounds in this process is still unresolved. The induction of antioxidant gene Cat1 found to take place independently both by ABA and ROS. As the H_2O_2 levels quickly enhance upon ABA treatment, H_2O_2 was projected to be a downstream component of ABA signaling associated with Cat1 induction (Guan et al. 2000). Therefore, a model for signaling pathway with ABA and ROS cross talking seems emerging. Briefly the process involves:

(a) ABA induced increase in cytosolic Ca^{2+} level (released from internal stores via ADP ribose and from external sources via influx across plasma membrane).
(b) ABA signaling pathway is negatively regulated by PP2C-like phosphatases that are transcriptionally activated by ABA, creating a negative feedback loop (Gosti et al. 1993).
(c) ABA induces generation of H_2O_2 in stomatal guard cells.
(d) H_2O_2 activated Ca^{2+} influx as well as stomatal closure.

In fact, H_2O_2 may adjust ABA responses by exerting the negative effect of PP2Cs on ABA pathway. As, for instance, a tobacco homolog of PP2Cs, implicated in ABA signaling, is transcriptionally downregulated by a number of oxidative stress stimuli (Varnova et al. 2000). Thus, any condition where ROS enhance (oxidative stress) downstream ABA activate the pathway by decreasing the negative regulator PP2C.

Oxidants, especially H_2O_2, cause Ca^{2+} influx into cytoplasm and consequently in mitochondria and nucleus. In fact, Ca^{2+} is recruited as signal molecule to respond oxidant stimuli or redox signal which can subsequently modulate the activities of proteins (transcription factors) and gene transcription. Usually redox signals in the form of oxidants upregulate Ca^{2+} channels or downregulate pumps and can even turn around Na^+/Ca^{2+} exchangers (Cunninghum and Fink 1996). However, the mechanisms by which the redox signals evoke such regulation are inadequately understood.

Calcineurin, a calcium-/calmodulin-activated serine/threonine phosphatase (PP2B), is a vital enzyme in Ca^{2+}-dependent eukaryotic signal transduction pathways. This class of PP2B plays a significant and decisive role in cellular responses to a variety of extracellular signals under unfavorable environmental cues, including stress adaptation (Asai et al. 1999; Krebs 1998). Its activity is found to be very sensitive to internal redox cue (Krebs 1998). H_2O_2 can completely abrogate calcineurin-mediated NF-AT transactivation in response to oxidative stress (Asai et al. 1999). It has also been revealed that calcineurin inhibitors defend against Ca^{2+}-induced cytotoxicity, which is observed when cells are transferred from Ca^{2+}-deficient media to Ca^{2+} containing media.

The stress hormone C_2H_4 also found to have interaction with ROS in their signaling networks. As evidence, it is found that tobacco plants with reduced peroxisomal catalase activity generate C_2H_4 as an earliest response under light stress

(Chamnongpol et al. 1998). Similarly, exogenous application of H_2O_2 augments C_2H_4 production in pine needles in a concentration-dependent manner (Ivenish and Tillberg 1995). Ozone induces accumulation of C_2H_4 along with ROS generation in tobacco plants. In all cases C_2H_4 seems to produce de novo, with concomitant production of ACC (precursor of C_2H_4 (Ivenish and Tillberg 1995)). Molecular genetic evidence with ethylene-insensitive *Arabidopsis* mutant ein-2 showed that O_3-induced ROS ($O_2^{\cdot -}$) accumulation instigates cellular senescence. On the contrary, ethylene-insensitive *Arabidopsis* mutant eto-1 is hypersensitive to ozone. This observation unequivocally suggests ROS-C_2H_4 interaction, where C_2H_4 has a potentiating role in redox-regulated cell death by controlling ROS titer.

All these evidences unambiguously pointed out the fact that plant growth regulators or hormones are not only positioned downstream of ROS signal, but ROS themselves are also signaling intermediate or second messenger in many hormone signaling pathways. And in most of the cases, Ca^{2+} can also be exploited as a signaling intermediate under oxidative stress. So, both feedback and feed-forward interactions may plausibly take place between different endogenous growth regulators, Ca^{2+} and ROS.

6.7 Lipid Peroxidation Products as "Biological Signals"

It is a matter of great surprise how a broad spectrum of environmental stress is being perceived and transduced to different kinds of nonspecific and specific response. One of the easiest ways to address this problem is to kick off a nonspecific fast response by transformation of cell membrane to signaling compounds with little energy expenditure. PUFA, being a most O_2-sensitive fatty acid and the one which are present in abundance in plant membrane, found to be the ideal compound to satisfy these conditions (Lynch and Thompson 1984). These membrane-associated PUFAs are stored in the exterior of the cell or organelle as free PUFA or may remain conjugated as phospholipids or galactolipid, which got hydrolyzed under elevated action of phospholipases under stress. Membrane lipid peroxidation (MLPO), both enzymatic and nonenzymatic, kicked off significantly under environmental stress with the generation of a host of secondary breakdown products, like alkoxy and peroxy radicals.

The alkoxy radicals generated by breakdown of 9-HPODE during MLPO is cleaved to 2, 4-decadenienal which can potentiate apoptosis (Miralto et al. 1999). The role of 2, 4-decadenienal in the induction of apoptosis in plants is of great significance (Grosch 1982). These secondary aldehydes represent the first product response of cells after oxidative modification of membrane architecture under stress. So, the formation of this class of secondary aldehydes in contrast to other bona fide stress signaling compounds like C_2H_4, ABA, JA, and SA does not essentially need a prior gene expression. Accordingly, as an originator of apoptotic signal, both linoleic acid and its oxidative decomposition product 2, 4-decadenienal (through MLPO) is functionally extremely significant. Comparable secondary oxidative products of MLPO are also supposed to be formed under oxidative stress, triggering

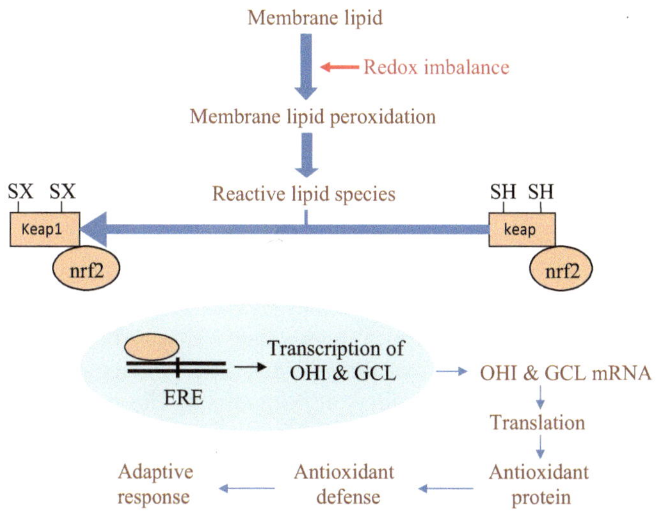

Fig. 6.7 Figure showing the mechanism of action of Reactive Lipid Species (RLS) derived from membrane lipid peroxidation (MLPO) on the expression of antioxidative defense genes or antioxidative response elements (Detail in text)

downstream signaling event (without gene expression), ultimately exhibiting a physiological response (Buchanon 1980; Stremler et al. 1989).

There are strong evidences to prove that the reactive lipid species (RLS), the products of MLPO cal, also contributes toward stress tolerance of plant. One such example of RLS-induced signaling is associated with the "protein sensor" keap1, which is an adaptor protein normally attached to transcript factor (TF) Nrf2. Oxidative stress-mediated electrophilic attack of RLS, particularly HNE to keap 1, causes release of the TF Nrf2 and subsequent translocation of transcription factor to the nucleus (Fig. 6.7). On migration to the nucleus, Nrf2 binds to its target ARE (antioxidative response element) and some other genes responsible for antioxidative defense proteins, like glutamate-cysteine ligase and AO-1, which are subsequently transcribed, upregulated, and ultimately causing the antioxidative defense response (Higden et al. 2012).

6.8 Conclusion and Perspective

The language of ROS signaling acts as one of the fundamental regulators of plant cell physiology and more particularly the cellular responses to environment. Apart from instigating cellular damage, ROS can act as ubiquitous "signaling molecule" in normal plant cell function. In fact, the level of ROS not only determine the redox status of the cell but also the responses, where at high concentrations, oxidative deterioration and cell death are initiated, or at low concentrations, ROS initiates expression of defense genes and adaptive responses. A complicated multicomponent redox circuit and complex regulatory system of antioxidant functions provide the flexibility necessary to allow the dual role of ROS or redox signal. But how this regulatory network functions and coordinated at cellular level to achieve the spatio-temporal level of ROS for initiating redox signaling is still poorly understood.

The complexity of regulation of ROS signaling resides primarily in its endogenous control of physiological titer and sensing. Studies till date largely centered on identifying mechanisms that regulate the function of ROS and their site-specific antioxidant-mediated breakdown. Subsequently, several studies also characterized many proteins sensitive to thiol oxidation and found to be relevant with its physiological role, particularly under unfavorable environmental condition. In fact, the significance of maintenance of endogenous titer of ROS like H_2O_2 is extremely important, and it is found that local antioxidant capacity determines the H_2O_2-mediated oxidative damage as signaling. In this aspect the regulatory role of antioxidant as H_2O_2 signaling molecule is also important. The studies described here largely centered around the sensing mechanisms of H_2O_2 under physiological conditions involving a wide array of redox sensors that primarily initiate cell signaling event. Studies described in this chapter suggest that low-concentration-mediated H_2O_2 signaling can be specific with specific molecular sensors with high degree of specificity. The study also revealed the fact that though H_2O_2 is a modulator of TFs and associated signaling, it also got the potential to be the regulator of expression of specific genes.

Despite remarkable progress in our understanding of ROS Biology, the exact character of ROS signaling network largely remains obscured. Both the molecular genetic approaches in addition to unique physiological study will be required to ascertain the exact position of ROS or redox signal in the elaborate signal transduction pathways. Further studies are also required to address the issues which will take into account the factors like sites of generation and spatiotemporal regulation of endogenous titer of ROS, the role of antioxidants and membrane architecture on the half-life of ROS, significance of pH and metal ion in sensing ROS, and protein-protein interaction of ROS sensors. These will ultimately help us immensely to unfold the unknown stories of ROS signaling.

References

Agrawal GK, Iwahashi H, Rakwal R (2003) Small GTPase 'Rop': molecular switch for plant defense responses. FEBS Lett 546:173–180

Ahn SG, Thiele DJ (2003) Redox regulation of mammalian heat shock factor 1 is essential for Hsp gene activation and protection from stress. Genes Dev 17:516–528

Allen JF (1993) Redox control of transcription: sensors, response regulators, activators and suppressors. FEBS Lett 332:203–207

Allen RG, Tresini M (2002) Oxidative stress and gene regulation. Free Radic Biol Med 28:463–499

Apel K, Hirt H (2004) Reactive oxygen species: metabolism, oxidative stress, and signal transduction. Annu Rev Plant Biol 55:373–399

Arrigo AP (1999) Gene expression and the thiol redox state. Free Radic Biol Med 27:936–945

Asai A, Qui J, Narita Y (1999) High level calcineurin activity predisposes neuronal cells to apoptosis. J Biol Chem 274:34450–34458

Balazadeh S, Wu A, Mueller-Roeber B (2010a) Salt-triggered expression of the ANAC092-dependent senescence regulon in *Arabidopsis thaliana*. Plant Signal Behav 5:733–735

Balazadeh S, Siddiqui H, Allu AD, Matallana-Ramirez LP, Caldana C, Mehrnia M, Zanor MI, Kohler B, Mueller-Roeber B (2010b) A gene regulatory network controlled by the NAC transcription factor ANAC092/AtNAC2/ORE1 during salt-promoted senescence. Plant J 62:250–264

Balazadeh S, Kwasniewski M, Caldana C, Mehrnia M, Zanor MI, Xue GP, Mueller- Roeber B (2011) ORS1, an H_2O_2-responsive NAC transcription factor, controls senescence in *Arabidopsis thaliana*. Mol Plant 4:346–360

Benabdellah K, Ruiz-Lozano JM, Aroca R (2009) Hydrogen peroxide effects on root hydraulic properties and plasma membrane aquaporin regulation in *Phaseolus vulgaris*. Plant Mol Biol 70:647–661

Bhattacharjee S (2005) Reactive oxygen species and oxidative burst: roles in stress, senescence and signal transduction in plants. Curr Sci 89:113–1121

Bhattacharjee S (2008) Calcium-dependent signaling pathway in heat-induced oxidative injury in *Amaranthus lividus*. Biol Plant 52:1137–1140

Bhattacharjee S (2010) Sites of generation and physicochemical basis of formation of reactive oxygen species in plant cell. In: Dutta Gupta S (ed) Reactive oxygen species and antioxidants in higher plants. Science Pubication. CRC Press, Boca Raton, pp 01–30

Bhattacharjee S (2012) An inductive pulse of hydrogen peroxide pretreatment restores redox-homeostasis andmitigates oxidative membrane damage under extremes of temperature in two rice cultivars (*Oryza sativa* L., Cultivars Ratna and SR 26B). Plant Growth Regul 68:395–410

Bienert GP, Møller ALB, Kristiansen KA, Schulz A, Moller IM, Schjoerring JK, Jahn TP (2007) Specific aquaporins facilitate the diffusion of hydrogen peroxide across membranes. J Biol Chem 282:1183–1192

Bolwell GP, Bnti VS, Davis DR, Zinmorlin A (1995) The origin of oxidative burst in plants. Free Radic Res 23:517–532

Buchanon BB (1980) Role of light in regulation of chloroplastic enzymes. Annu Rev Plant Physiol 31:341–374

Chen Z, Silva H, Klessig DF (1993a) Active oxygen species in the induction of plant systemic acquired resistance by salicylic acid. Science 262:1883–1886

Chen Z, Silva H, Klessig DF (1993b) Active oxygen species in the induction of plant systemic acquired resistance by salicylic acid. Science 262:1886–1891

Chen W, Chao G, Singh KB (1996) The promoter of a H_2O_2-inducible, *Arabidopsis* glutathione S-transferase gene contains closely linked OBF- and OBP1-binding sites. Plant J 10:955–963

Chowdhury FK, Rivero FM, Blumwald E, Mittler R (2016) Reactive oxygen species, abiotic stress and stress combination. Plant J. https://doi.org/10.1111/tpj.13299

Ciftci-Yilmaz S, Morsy MR, Song L, Coutu A, Krizek BA, Lewis MW, Warren D, Cushman J, Connolly EL, Mittler R (2007) The EAR-motif of the Cys2/His2-type zinc finger protein Zat7 plays a key role in the defense response of *Arabidopsis* to salinity stress. J Biol Chem 282:9260–9268

Chamnongpol S, Willekens H, Moeder W, Langebartels C, Sandermann H Jr, Van Montagu M, Inzé D, Van Camp W (1998) Defense activation and enhanced pathogen tolerance induced by H_2O_2 in transgenic plants. Proc Natl Acad Sci U S A 95:5818–5823

Cunninghum KW, Fink GK (1996) Calcineurin-dependent inhibits VCX1-dependent H+/Ca2+ exchange and induces Ca2+-ATPases in *Saccharomyces cerevisiae*. Mol Cell Biol 16:2226–2237

Daniel V (1993) Glutathione S-transferases: gene structure and regulation of expression. Crit Rev Biochem Mol Biol 28:173–207

Davletova S, Rizhsky L, Liang H, Shengqiang Z, Oliver DJ, Coutu J, Shulaev V, Schlauch K, Mittler R (2005) Cytosolic ascorbate peroxidase 1 is a central component of the reactive oxygen gene network of *Arabidopsis*. Plant Cell 17:268–281

Demple B (1991) Regulation of bacterial oxidative stress genes. Annu Rev Gen 25:315–337

Dröge-Laser W, Kaiser A, Lindsay WP, Halkier BA, Loake GJ, Doerner P, Dixon RA, Lamb C (1997) Rapid stimulation of a soybean protein-serine kinase that phosphorylates a novel bZip DNA-binding protein, G/HBF-1, during the induction of early transcription-dependent defenses. EMBO J 16:726 734

Dubbs JM, Mongkolsuk S (2012) Peroxide-sensing transcriptional regulators in bacteria. J Bacteriol 194:5495–5503

Dynowski M, Schaaf G, Loque D, Moran O, Ludewig U (2008) Plant plasma membrane water channels conduct the signalling molecule H_2O_2. Biochem J 414:53–61

Eggler AL, Liu G, Pezzuto JM, van Breemen RB, Mesecar AD (2005) Modifying specific cysteines of the electrophile-sensing human Keap1 protein is insufficient to disrupt binding to the Nrf2 domain Neh2. Proc Natl Acad Sci U S A 102:10070–10075

Eulgem T, Somssich IE (2007) Networks of WRKY transcription factors in defense signalling. Curr Opin Plant Biol 10:366–371

Fernandes L, Rodrigues-Pousada C, Struhl K (1997) Yap, a novel family of eight bZIP proteins in *Saccharomyces cerevisiae* with distinct biological functions. Mol Cell Biol 17:6982–6993

Ferrer-Sueta G, Manta B, Botti H, Radi R, Trujillo M, Denicola A (2011) Factors affecting protein thiol reactivity and specificity in peroxide reduction. Chem Res Toxicol 24:434–450

Foreman J, Demidchik V, Bothwell JH, Mylona P, Miedema H, Torres MA, Linstead P, Costa S, Brownlee C, Jones JD, Davies JM, Dolan L (2003) Reactive oxygen species produced by NADPH oxidase regulate plant cell growth. Nature 422:442–446

Friling RS, Bergelson S, Daniel V (1992) Two adjacent AP-1-like binding sites from the electrophile-responsive element of the murine glutathione S-transferase Ya subunit gene. Proc Natl Acad Sci U S A 89:668–673

Gadjev I, Vanderauwera S, Gechev TS, Laloi C, Minkov IN, Shulaev V, Apel K, Inze D, Mittler R, Van Breusegem F (2006) Transcriptomic footprints disclose specificity of reactive oxygen species signalling in Arabidopsis. Plant Physiol 141:436–445

Gechev TS, Hille J (2005) Hydrogen peroxide as a signal controlling plant programmed cell death. J Cell Biol 8:17–20

Gechev TS, Minkov IN, Hille J (2005) Hydrogen peroxide-induced cell death in *Arabidopsis*: transcriptional and mutant analysis reveals a role of an oxoglutarate-dependent dioxygenase gene in the cell death process. Int Union Biochem Mol Biol Life 57:181–188

Ghesquière B, Jonckheere V, Colaert N, Van Durme J, Timmerman E, Goethals M, Schymkowitz J, Rousseau F, Vandekerckhove J, Gevaert K (2011) Redox proteomics of protein-bound methionine oxidation. Mol Cell Proteomics 10:M110.006866

Gill SS, Tuteja N (2010) Reactive oxygen species and antioxidant machinery in abiotic stress tolerance in crop plants. Plant Physiol Biochem 48:909–930

Godon C, Lagniel G, Lee J, Buhler JM, Kieffer S, Perrot M, Boucherie H, Toledano MB, Labarre J (1998) The H_2O_2 stimulon in *Saccharomyces cerevisiae*. J Biol Chem 273:224–280

Gong M, Li YJ, Chen SZ (1998) ABA induced thermotolerance in maize seedlings is mediated by calcium and associated antioxidant systems. J Plant Physiol 153:488–497

Gosti F, Beandoin N, Serizet C, Webb AAR, Vertanian N, Giraudat T (1993) AB11 protein phosphatase 2C is a negative regulator of ABA signaling. Plant Cell 11:1897–1903

Grant M, Brown I, Adams S, Knight M, Ainslie A, Mansfield J (2000) The *RPM1* plant disease resistance gene facilitates a rapid and sustained increase in cytosolic calcium that is necessary for the oxidative burst and hypersensitive cell death. Plant J 23:441–450

Grosch W (1982) Lipid degradation products and flavor. In: Morton ID, Mac Leod AJ (eds) Food flavours. Elsevier, Amsterdam, pp 325–398

Guan ZM, Zhao J, Scandalios JG (2000) Cis-elements and trans-factors that regulate expression of maize Cat1 antioxidant gene in response to ABA and osmotic stress: H_2O_2 is the likely intermediary signaling molecule for the response. Plant J 22:87–98

Halliwell B (2006) Reactive species and antioxidants. Redox biology is a fundamental theme of aerobic life. Plant Physiol 141:312–322

Helmann JD, Wu MF, Gaballa A, Kobel PA, Morshedi MM, Fawcett P, Paddon C (2003) The global transcriptional response of *Bacillus subtilis* to peroxide stress is coordinated by three transcription factors. J Bacteriol 185:243–253

Higden A, Olver AR, Jyob J, Lender A, Darlayusmer VM (2012) Cell signalling by reactive lipid species: new concept and molecular mechanism. Biochem J 442:453–464

Ivenish G, Tillberg EJ (1995) Stress-induced ethylene biosynthesis in pine needles: a search for putative ACC- independent pathway. Plant Physiol 145:308–317

Karimpour S, Lou J, Lin LL, Rene LM, Langunas L, Ma X, Karra S, Bradbarey CM, Markovina S, Goswami PC et al (2002) Thioredoxin reductase regulates AP-1 activity as well as thioredoxin nuclear localization via active cysteines in response to ionizing radiation. Oncogene 21:6317–6327

Krebs J (1998) The role of calcium in apoptosis. Biometals 11:375–382

Kim SO, Merchant K, Nudelman R, Beyer WF Jr, Keng T, DeAngelo J, Hausladen A, Stamler JS (2002) OxyR: a molecular code for redox-related signaling. Cell 109:383–396

Kovalchuk I (2010) Multiple roles of radicals in plants. In: Dutta Gupta S (ed) Reactive oxygen species and antioxidants in higher plants. Science Pubication. CRC Press, Boca Raton, pp 31–44

Kurepa J, Herooart D, Van Motague M, Inze D (1997) Differential expression of cu, Zn and Fe-superoxidedismutase genes of tobacco during development, oxidative stress and hormonal treatments. Plant Cell Physiol 38:463–473

Lachaud C, Da Silva D, Amelot N, Beziat C, Briere C, Cotelle V, Graziana A, Grat S, Mazars C, Thuleau P (2011) Dihydrosphingosine-induced programmed cell death in tobacco BY-2 cells is independent of H_2O_2 production. Mol Plant 4:310–318

Lamb C, Dixon RA (1997) The oxidative burst in plant disease resistance. Annu Rev Plant Physiol Plant Mol Biol 48:251–275

Lee C, Lee SM, Mukhopadhyay P, Kim SJ, Lee SC, Ahn WS, Yu MH, Storz G, Ryu SE (2004) Redox regulation of OxyR requires specific disulfide bond formation involving a rapid kinetic reaction path. Nat Struct Mol Biol 11:1179–1185

Leon J, Lawton MA, Raskin I (1995) Hydrogen peroxide stimulates salicylic acid biosynthesis in tobacco. Plant Physiol 108:1673–1679

Li T, Li H, Zhang YX, Liu JY (2011) Identification and analysis of seven H_2O_2-responsive miRNAs and 32 new miRNAs in the seedlings of rice (*Oryza sativa* L. ssp. *indica*). Nucleic Acids Res 39:2821–2833

Locato V, Pinto de MC, Paradiso A, Gara de L (2010) Reactive oxygen species and ascorbate – glutathione interplay in signaling and stress responses. In: Dutta Gupta S (ed) Reactive oxygen species and antioxidants in higher plants. Science Pubication. CRC Press, Boca Raton, pp 45–64

Lynch DV, Thompson JE (1984) Lipoxygenase mediated production of superoxide anion in senescing plant tissue. FEBS Lett 173:251–254

Mahalingam R, Fedoroff N (2003) Stress response, cell death and signaling. Physiol Plant 119:56–68

Marinho HS, Real C, Cyrne L, Soares H, Antunes F (2014) Hydrogen peroxide sensing, signaling and regulation of transcription factors. Redox Biol 2:535–562

Maxwell DP, Wang Y, McIntosh L (1999) Alternative oxidase lowers mitochondrial ROS production in plant cells. Proc Natl Acad Sci U S A 96:8271–8276

Miao Y, Laun T, Zimmermann P, Zentgraf U (2004) Targets of the WRKY53 transcription factor and its role during leaf senescence in *Arabidopsis*. Plant Mol Biol 55:853–867

Miller G, Shulaev V, Mittler R (2008) Reactive oxygen signalling and abiotic stress. Physiol Plant 133:481–489

Miller G, Schlauch K, Tam R, Cortes D, Torres MA (2009) The plant NADPH oxidase RbohD mediates rapid, systemic signaling in response to diverse stimuli. Sci Signal 2(84):45–49

Miller G, Suzuki N, Ciftci-Yilmaz S, Mittler R (2010) Reactive oxygen species homeostasis and signalling during drought and salinity stresses. Plant Cell Environ 33:453–467

Miralto A, Barone G, Romano G, Poulet SA, Lanora A, Russo GL, Buttino I, Mazarella G, Laabir M, Cabrini M, Giacobbe MG (1999) The insidious effect of diatoms on copepod reproduction. Nature 402:173–176

Mittler R, Vanderauwera S, Suzuki N, Miller G, Tognetti VB, Vandepoele K, Gollery M, Shulaev V, Van Breusegem F (2011) ROS signaling: the new wave? Trends Plant Sci 16:300–309

Mubarakshina MM, Ivanov BN, Naydov IA, Hillier W, Badger MR, Krieger-Liszkay A (2010) Production and diffusion of chloroplastic H_2O_2 and its implication to signalling. J Exp Bot 61:3577–3587

Mylona PV, Polidoros AN (2010) ROS regulation and antioxidant genes. In: Dutta Gupta S (ed) Reactive oxygen species and antioxidants in higher plants. Science Pubication. CRC Press, Boca Raton, pp 47–59

Neill S, Desikan R, Clarke A, Harcock J (1999) H_2O_2 signaling in plant cells. In: Smallwood MF, Calvert CM, Bowles DJ (eds) Plant responses to environmental stresses. Bios Sci Pub, Oxford, pp 59–64

Nishiyama A, Masutani H, Nakamura H, Nishinaka Y, Yodi J (2001) Redox regulation by thioredoxin and thioredoxin binding proteins. IUBMB Life 52:29–33

Noctor G, Foyer CH (1999) Ascorbate and glutathione:keeping active oxygen species under control. Annu Rev Plant Physiol Mol Biol 49:249–279

Pei ZM, Murata Y, Benning G, Thomine S, Klusener B, Allen GJ, Grill E, Schroeder JI (2000) Calcium channels activated by hydrogen peroxide mediate abscisic acid signalling in guard cells. Nature 406:731–734

Petrov VD, Van Breusegem F (2012) Hydrogen peroxide-a central hub for information flow in plant cells. AoB Plants:pls014. https://doi.org/10.1093/aobpla/pls014

Polidoros AN, Scandalios JG (1999) Role of hydrogen peroxide and different classes of antioxidants in the regulation of catalase and glutathione S-transferase gene expression in maize (*Zea mays* L.). Physiol Plant 106:112–119

Rentel MC, Knight MR (2004) Oxidative stress-induced calcium signalling in Arabidopsis. Plant Physiol 135:1471–1479

Rizhsky L, Liang H, Mittler R (2003) The water-water cycle is essential for chloroplast protection in the absence of stress. J Biol Chem 278:38921–38925

Robson CA, Vanlerberghe GC (2002) Transgenic plant cells lacking mitochondrial alternative oxidase have increased susceptibility to mitochondria –dependent and independent pathways of cell death. Plant Physiol 129:1908–1920

Rouhier N, Lamaire SD, Jaqcquot JP (2009) The role of GSH in photosynthetic organisms: the emerging function of glutaredoxins and glutathionylation. Annu Rev Plant Biol 59:143–166

Sakamoto M (2008) Involvement of hyrogen peroxide in leaf abscission signaling revealed by analysis with an in vitro abscission system in Capsicum plants. Plant J 56:13–27

Schell MA (1993) Molecular biology of the LysR family of transcriptional regulators. Annu Rev Microbiol 47:597–626

Stephen DWS, Rivers SL, Jamieson DJ (1995) The role of the YAP1 and YAP2 genes in the regulation of the adaptive oxidative stress responses of *Saccharomyces cerevisiae*. Mol Microbiol 16:415 423

Storz G, Tartaglia LA, Ames BN (1990) Transcriptional regulator of oxidative stress-inducible genes: direct activation by oxidation. Science (New York, NY) 248:189–194

Stremler KE, Staforrini DM, Prescott SM, Zimmerman GA, Mcintyre TM (1989) An oxidized derivative of phosphatidylcholine is a substrate for the platelet activating factor acetylhydrolase from human plasma. J Biol Chem 264:5331–5334

Tonks NK (2005) Redox redux: revisiting PTPs and the control of cell signaling. Cell 121:667–670

vanMontfort RL, Congreve M, Tisi D, Carr R, Jhoti H (2003) Oxidation state of the active-site cysteine in protein tyrosine phosphatase 1B. Nature 423:773–777

Varnova E, Langebartels C, Van Montague M, Inze D, Van Camp W (2000) Oxidative stress, heat shock and drought differentially affect expression of a tobacco protein phosphatase 2C. J Exp Bot 51:1763–1775

Varnova E, Inzé D, Van Breusegem F (2002) Signal transduction during oxidative stress. J Exp Bot 53:1227–1236

Veal EA, Day AM, Morgan BA (2007) Hydrogen peroxide sensing and signaling. Mol Cell 26:1–14

Wingate VPM, Lawton MA, Lamb CJ (1988) Glutathione causes a massive and selective induction of plant defense genes. Plant Physiol 87:206–211

Wu J, Shang Z, Wu J, Jiang X, Moschou PN, Sun W, Roubelakis-Angelakis KA, Zhang S (2010) Spermidine oxidase-derived H_2O_2 regulates pollen plasma membrane hyperpolarization-activated Ca(2+) -permeable channels and pollen tube growth. Plant J 63:1042–1053

Yamamoto A, Mizukami Y, Sakurai H (2005) Identification of a novel class of target genes and a novel type of binding sequence of heat shock transcription factor in *Saccharomyces cerevisiae*. J Biol Chem 280:11911–11919

Yamamoto A, Ueda J, Yamamoto N, Hashikawa N, Sakurai H (2007) Role of heat shock transcription factor in *Saccharomyces cerevisiae* oxidative stress response. Eukaryot Cell 6:1373–1379

You J, Chan J (2015) ROS regulation during abiotic stress responses in crop plants. Front Plant Sci. https://doi.org/10.3389/fpls.2015.01092

Zheng M, Aslund F, Storz G (1998) Activation of the OxyR transcription factor by reversible disulfide bond formation. Science 279:1718–1721

Exploring Oxidative Stress in Plants: Proteomic and Genomic Approaches

7

Abstract

Formation of reactive oxygen species (ROS), maintenance of their titer through antioxidative defense, and subsequent redox regulation are inescapable events in plant cell during normal metabolic processes. Onset of environmental stresses, both abiotic and biotic, increase the intensity of ROS production many times in plant cell, which leads to a situation, where cell loose redox homeostasis oxidative stress. ROS, particularly hydrogen peroxide (H_2O_2), hydroxyl radicals ($OH^{\bullet}$), and superoxide radicals (O_2^-), are capable of causing serious damage to all important cellular metabolites and macromolecules like lipids, carbohydrates, proteins, and nucleic acids. To deal with a situation of oxidative stress, plants possess competent antioxidant defense machinery, comprising of antioxidative enzymes, water- and lipid-soluble antioxidant molecules, and ROS quenchers. The equilibrium between oxidative load and antioxidant defense capacity of plant cell decides the fate of the plant. Furthermore and more surprisingly, plants have also evolved a way to exploit lower titer of ROS as signaling component to regulate wide variety of plant processes like cell elongation, differentiation, morphogenesis, responses to environmental stress, etc. Therefore, to understand and explore the oxidative stress situation, accurate assessment of oxidative stress in plants requires sensitive and robust assays for both detection and quantification of ROS in the cell, as well as the identification and characterization of oxidative stress-induced changes at genomic and proteomic levels. Employing these approaches, not only the potential role played by ROS-induced modifications in stressed cells can be deciphered but also will help us to explore the ROS biology for crop improvement. This chapter presents an outline of the latest research endeavors in the field of ROS biology employing plant proteomics and genomics aimed at identifying the role of transcriptional and posttranslational modifications in plant cell under oxidative stress.

Keywords

Oxidative stress · Omic sciences · Reactive oxygen species · Posttranslational oxidative modification of proteins

S. Bhattacharjee, *Reactive Oxygen Species in Plant Biology*,
https://doi.org/10.1007/978-81-322-3941-3_7

7.1 Introduction

Environmental stresses cause significant decreases in the crop growth and productivity mainly via oxidative stress, which occurs due to redox imbalance. In fact, ROS are unwelcome by-product of aerobic metabolism and constantly produced during redox metabolisms in the cell (Foyer and Noctor 2005). However, the redox homeostasis is tightly controlled by redundant antioxidative protective mechanism (Gill and Tuteja 2010). ROS also found to have important regulatory and signaling properties in cellular physiology (Foyer and Noctor 2005; Gill and Tuteja 2010). Moreover, the origin of oxidative stress triggered by hyper-accumulation of ROS [like, hydroxyl radicals ($OH^•$), hydrogen peroxide (H_2O_2), superoxide radicals ($O_2^{\cdot -}$), peroxy and alkoxy radicals ($RO^•$, $RCO^•$)] reactive nitrogen-oxygen species (RNS) [like, nitric oxide ($NO^•$), peroxynitrite ($ONOO^-$), etc.] escapes antioxidant defense machinery and may evoke metabolic dysfunction. Among biomolecules, proteins are the major target of ROS that constitute about one third of the oxidized species. Since ROS are highly toxic, reactive, and extremely short-lived, it is very difficult to quantify them directly. This can be done by following indirect approach by measuring different components oxidized by these species that requires the use of sensitive, robust, and sophisticated techniques. The ROS-lead oxidative stress at genomic level is another area of investigation to understand plant responses to unfavorable environmental cues.

The proteomic studies in relation to environmental stress-caused oxidative stress, like identification of carbonyl groups in proteins, mainly depend on classical biochemical approaches. The immunological detection of modified protein residues under oxidative stress has also been a widely used method for the identification of the stress-induced modifications on proteins (Levine et al. 1994). However, all of these assays have great disadvantages in terms of large number of artifacts during sample preparation and can't differentiate between different pro-oxidant-mediated damages (Buss et al. 1997; Cuddihy et al. 2008). Moreover, focusing on a single oxidative modification does not render the complexity of the cellular response to oxidative stress, or regulation of redox protein. The cellular response to oxidative stress is extremely dynamic that merely results in the simple accumulation of modified proteins. Therefore, a classical approach to the event based on limited information will not reflect the complexity of redox response, which involves a variety of changes in protein levels (both reversible and irreversible), controlled posttranslational modifications, and oxidative damage to proteins (Rabilloud et al. 2005). In this perspective, it has been found that proteomics is the best suited approach to this problem. The improvement in the sophisticated analytical techniques like mass spectrometry has by now provided more precise and more quantitative ways to measure oxidative modifications in the cell. A new branch of proteomics is called "redox proteomics," in which posttranslational modifications of proteins oxidative stress can be studied. This can be used in studying the proteins showing alterations under varied magnitude of oxidative stress (Moller et al. 2007). Considering the significance of detection of genes for plant oxidative stress tolerance, it has been considered indispensable to get insights into the genetics of stress acclimation which will enable us ultimately to successfully develop transgenic plants to be grown under unfavorable environmental cues. With the help of identification of the stress tolerance gene and their expression patterns, it has become possible to bring them under hybridization program or transgenesis for making oxidative stress-tolerant plants (Gill and Tuteja 2010).

7.2 Different Proteomic Approaches Used for the Identification of ROS-Modified Proteins

Several proteomic tools are in vogue for the detection of oxidatively modified proteins or identification of stress-responsive and/or stress-induced proteins and the analysis of their differential expression under stress. The list of different proteomic approaches or tools used for the analysis of redox modification of proteins has been summarized in Fig. 7.1, and the same have been discussed. In fact, under normal

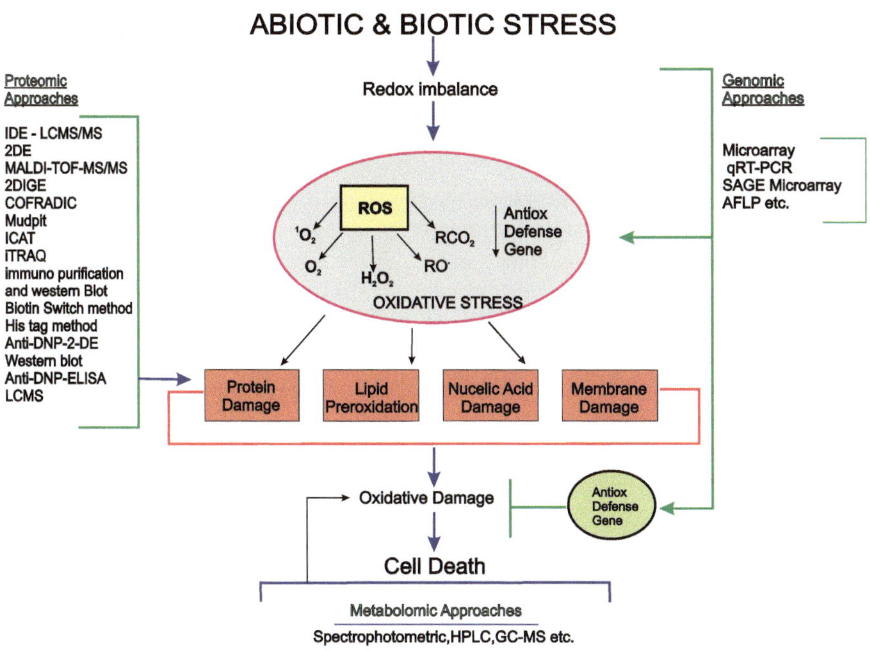

Fig. 7.1 Figure representing a list of different proteomic, genomic and metabolomic approaches used to detect oxidative stress-responsive changes in plants

metabolism in plants, redox reactions take place in chloroplast, mitochondria, peroxisome, and plasma membrane which produces ROS by leakage of electrons onto molecular oxygen or by oxidoreductase enzymes which subsequently being scavenged by antioxidant defense mechanisms. However, under environmental stress, both abiotic and biotic, the redox homeostasis is disturbed, which is governed by elevated rate of ROS production and lesser elimination by antioxidant system. The rise in ROS in cells is sensed by the plant cells and transduced in the entire system by different signaling pathways for stress acclimation. The study of redox proteome helps us to reveal the route for signaling pathway during oxidative stress. ROS reversibly or irreversibly modify amino acid residues in oxidation-susceptible proteins. Cysteine, methionine, and selenocysteine undergo reversible oxidation, while tryptophan, tyrosine, arginine, histidine, phenylalanine, cysteine, valine, leucine, isoleucine, lysine, etc. undergo irreversible oxidation. These oxidative modifications alter the structure of proteins which is sensed by the plant and thus initiate signaling cascade.

During oxidative stress, the amino acid residues of ROS vulnerable proteins are oxidized either directly by ROS or indirectly by reaction with secondary by-products of oxidative stress (MLPO, PO products). The best-studied proteins that have undergone oxidative posttranslational modifications are those enzymes involved in the Calvin cycle (Schurmann and Buchanan 2008), sulfur metabolism (Kopriva et al. 2012), and starch metabolism (Glaring et al. 2012).The thiol of Cys is the most extensively characterized component of the redox proteome as Cys side chain contains sulfur atom at the core of the thiol which is electron-rich and its d-orbitals allow for multiple oxidation states. However, not all Cys residues in a protein are prone to ROS-mediated modifications, and the reactivity of different thiol-proteins toward ROS varies according to their physiological function and local redox environment. In individual Cys residues, the reactivity is strongly correlated with their pKa, i.e., the ability to form the anionic form of the sulfur, called thiolate (R-S–), which is much more reactive than the thiol. If the pKa of the sulfur atom is higher than the pH of the solution, the protonated thiol will be the dominant species. However, if the pKa is lower than the pH, the majority of the thiols will be present as a thiolate (Cys prone to oxidation). The pKa of Cys residues is largely determined by the local electrostatic environment, i.e., the presence of proximal charged residues or dipoles and the hydrogen bonding between thiols/thiolates and neighboring residues (Harris and Turner 2002). Thus cysteine oxidation permits various posttranslational modification resulting in diverse regulatory effects. When cellular oxidative strength is low, reversible oxidation of cysteine residue to sulfenic acid takes place; this modification is highly unstable and leads to further modifications (Claiborne et al. 1993). An excess concentration of oxidant can lead to further oxidation to sulfenic acid (R-SO2H) and subsequently to irreversible sulfonic acid (R-SO3H; Roos and Messens 2011). Alternatively, sulfenic acid can react with free protein thiols to form intra- or intermolecular disulfide bonds (R-S-S-R/R-S-S-R′) or is modified by low molecular weight thiols (like GSH in plants), leading to Cys S-glutathionylation.

Similarly, oxidation of methionine residue under biological conditions leads to formation of methionine sulfoxide. Further round of oxidation leads to the formation of sulfone which is irreversible reaction and is damaging to the proteins. The restoration of methionine from methionine sulfoxide is catalyzed by methionine sulfoxide reductase using thioredoxin as a reductant. Therefore, methionine can act as a an antioxidant to protect other amino acid residue from oxidative damage (Levine et al. 1996). Protein cysteine residues can also undergo reversible S-glutathionylation via mixed disulfide formation under oxidative stress conditions (Xiong et al. 2011; Pastore and Piemonte 2012; Pimentel et al. 2012). Glutathione (GSH) is a cellular antioxidant and oxidizes cysteine residue so that there is no overoxidation and permanent damage to redox proteomes. Glutathione modification of proteins also affects their functional role (Zweier et al. 2011). The modified protein impacts gene expression, cell proliferation, and apoptosis protecting the cellular machinery from oxidative insults (Xiong et al. 2011; Zweier et al. 2011).

Protein carbonyls are formed from several amino acids residues including arginine, histidine, lysine, proline, threonine, and cysteine under oxidative stress. And there is formation of reactive aldehyde or ketone groups causing inactivation, cross-linking, or breakdown of proteins (Levine et al. 1996). Carbonylation occurs mainly via the HO$^{\bullet}$ radical, and this has been suggested to be rather nonspecific (Levine and Stadtman 2001). During oxidative stress, oxidation of protein carbonyl group is very common. And its magnitude is much greater than any other modifications that occur only on a specific amino acid residue (Yan 2009; Yan and Forster 2011) and thus is more readily detectable. Therefore, it can well be used as biomarker for measurement of protein oxidation and oxidative stress under environmental stress, aging, and infection (Chakraborty and Bhattacharjee 2015; Stadtman 1992, 2001, 2006; Stadtman et al. 2003; Yan 2009; Yan and Sohal 1998, 2000; Yan et al. 1997, 1998, 2002; Groeger et al. 2009; Knoefler et al. 2012; Madian and Regnier 2010; Wehr and Levine 2013).

Indirect oxidative modification of protein amino acid side chains occurs through the formation of adducts with products of oxidatively modified lipids (lipid radicals), amino acids, sugars, and glutathione. For example, lipid peroxidation breakdown products such as hydroxynonenal (HNE), malondialdehyde (MDA), and acrolein bind covalently to Lys, His, and Cys residues, leading to the addition of aldehyde moieties to the protein. Carboxymethyllysine represents a form of protein modification generated by the oxidation products of sugars.

7.2.1 Proteomic Approach for Molecular Characterization of Oxidized Protein Residues

To identify and quantify the oxidative modifications mentioned above, several proteomic technologies are exploited. New experimental approaches have been developed as well to identify and characterize specific thiol modifications and/or changes under different redox status of plant tissues. Both indirect and direct methods have been developed to investigate oxidative modification of cysteine. The majority of indirect methods to monitor changes in the redox state of cysteines rely on the loss of reactivity with thiol modifying reagents or restoration of labeling by reducing agents. With the advancement of chemical biology, methods are developed where small molecular probes made to react selectively with specific cysteine oxoform modification. However, the key challenge is to design such probe. Methods existing at present involve quantification of reactive cysteines with alkylated free thiol followed by various detection methods to monitor labeled proteins. N-Ethylmaleimide (NEM) 1 and iodoacetamide (IAM) 2 are both thiol-specific alkylating agents that react with cysteine thiolate anions more readily than with free thiols (Ying et al. 2007). The reaction of NEM with thiols is more rapid as compared to IAM; however, in some cases, it may also be less specific. Both reagents form covalent bonds with reactive cysteines and can be derivatized with biotin, fluorophores, and stable and radioactive isotopes. Biotin can be used as a handle to facilitate Western blot detection or enrichment using streptavidin-conjugated reagents, while radiolabels or fluorophores enable the detection of modified cysteines by one-dimensional or two-dimensional gel electrophoresis.

Leichert et al. (2008) have developed the OxICAT method to quantify reversible oxidation of cysteine residues. The OxICAT approach couples ICAT with differential thiol trapping that involves five key steps:

1. Cell lysis in the presence of trichloroacetic acid to inhibit thiol/disulfide exchange.
2. Protein denaturation and alkylation with the light (12C) ICAT reagent.
3. Treatment with tris(2-carboxyethyl)phosphine (TCEP), which returns reversibly oxidized cysteine residues (i.e., disulfides, nitrosothiols, sulfenic acids) to their thiol form.
4. Alkylation of nascent free thiols with the heavy (13C) ICAT reagent.
5. The ratio of light to heavy isotope-labeled cysteines is quantified by LC-MS.

However, the only direct and high-throughput method to identify proteins that undergo oxidant-induced disulfide formation is sequential nonreducing/reducing two-dimensional gel electrophoresis, also known as Redox 2D-PAGE (Cumming 2008). In this procedure, proteins are first electrophoresed under nonreducing conditions. In subsequent steps, proteins are reduced in gel, excised, layered onto a second gel, and electrophoresed. Proteins that do not contain disulfide bonds will migrate on the diagonal across the gel in this system, whereas proteins with intermolecular or intramolecular disulfide bonds migrate below or above the diagonal, respectively. After separation by 2D-PAGE, oxidized proteins can be cut from the gel and identified by LC-MS. The greatest limitation of this technique is that it cannot reliably visualize or produce analytical quantities of proteins that are present in less than 1000 copies per cell. Nonetheless, this strategy has successfully identified cysteines in more abundant proteins that are susceptible to intermolecular and intramolecular disulfide formation in cells (Cumming 2008).

Current methods for measurement of methionine sulfoxide for in vitro samples employ amino acid analysis following CNBr cleavage (Maier et al. 1995; Levine et al. 1996). To date, there are no antibodies available which can specifically detect Met sulfoxide, although some laboratories have tried to generate them. It is not clear whether Met sulfoxide is immunogenic.

Highly sensitive assays are available for detection of carbonyl group of proteins, produced under oxidative stress. Many of the assays involve derivatization of the carbonyl group with dinitrophenylhydrazine (DNPH), which leads to formation of a stable dinitrophenylhydrazone product (Levine et al. 1990). This then can be detected by various means. The DNP group itself absorbs ultraviolet (UV) light at 370 nm, with a molar extinction coefficient of 22,000, so that the total carbonyl content of a protein or mixture of proteins can be quantified by the ratio of the absorbance at 370 nm to that at 280 nm (Levine et al. 1990). The spectrophotometric DNPH assay can be coupled to protein fractionation by high-performance liquid chromatography (HPLC) to give greater sensitivity and specificity than measuring total carbonyls in a protein mixture (Levine et al. 1994).

Other DNPH-associated assays utilize anti-DNP antibodies to quantify carbonyl content. Total carbonyls can be measured by enzyme-linked immunosorbent assay (ELISA) (Buss et al. 1997), slot blotting (Robinson et al. 1999), or immunohistochemistry (Smith et al. 1998); the carbonyl content in individual proteins is assessed by one-dimensional (Shacter et al. 1994; Keller et al. 1993) or two-dimensional (Nakamura and Goto 1996) sodium dodecyl sulfate (SDS) gel electrophoresis followed by immunoblotting (Western blotting). These latter two methods have significantly more sensitivity and specificity than all other total carbonyl assays.

Carbonyls are relatively difficult to be oxidized compared to Met sulfoxide and cysteinyl derivatives or other secondary products of oxidative deterioration and may thus be reflective of more severe cases of oxidative stress and a sensitive biomarker for the same. Indeed, detection of elevated levels of protein carbonyls is generally a sign not only of oxidative stress but also of a stress-associated dysfunction. On the other hand, although Met and Cys are the two amino acyl residues that are most highly susceptible to oxidative attack, modification of these residues may not always result in an effect on protein function (Reddy et al. 1994; Gitlin et al. 1996). In fact, methionine and cysteine residues may, in some cases, serve as internal scavengers which protect critical amino acid residues from oxidative assaults (Levine et al. 1996).

To identify cysteine nitrosylation, anti-*S*-nitrosocysteine antibodies have been generated and applied in immunohistochemistry studies. Recently, the biotin switch method was established to allow detection of protein S-nitrosylation after SDS-polyacrylamide gel electrophoresis and electroblotting (Jaffrey and Snyder 2001). First, free thiols are blocked by incubation with a thiol-specific methiolating agent (methyl methanethiosulfonate or MMTS) in the presence of sodium dodecyl sulfate to ensure access to buried cysteine residues. Conversely, the methiolating agent does not react with nitrosothiols or disulfide bonds under the reaction conditions used. Excess of MMTS is removed using a spin column or by acetone precipitation. Next, nitrosothiol bonds are selectively decomposed to thiols with ascorbate. Finally, the free thiol groups are reacted with the sulfhydryl-specific biotinylation reagent, *N*-[6-(biotinamido)hexyl]-3′-(2′-pyridyldithio) propionamide (biotin-HPDP). The labeled proteins may then be separated by SDS-polyacrylamide gel electrophoresis, electroblotted, and detected by standard immunoblotting methods using anti-biotin antibodies or streptavidin. Labeled proteins may be selectively enriched using streptavidin affinity chromatography as well (Sell et al. 2008).

7.2.2 Two-Dimensional Electrophoreses Coupled with MALDI TOF MS/MS

Most of the oxidative modifications in proteins under environmental stress are quantitative and deals with posttranslational modifications which necessitate the use of the right approach for the proteome study. In addition, no biochemical tools are on hand that can address all recognized oxidative modifications. These problems rule out certain features of proteomics, such as isotope-coded, peptide-based methods (Gygi et al. 1999) and shotgun approaches (Washburn et al. 2001) for analysis. Till date, most of the proteomic studies have used 2DE as a protein separation and quantification tool and mass spectrometry (MS) as a protein characterization tool in response to oxidative stress. This technique is able to separate proteins through quite variations in the isoelectric point (pI), and many oxidative modifications pertaining to posttranslational changes can be detected in this way. In *Oryza sativa* embryos (Fukuda et al. 2003), roots (Yan et al. 2005), leaves (Salekdeh et al. 2002), and anthers (Liu and Bennett 2011), stress-induced proteins were identified and characterized through proteomic analysis like 2DE coupled with MS. Cold acclimation associated proteins in wheat were analyzed through 2DE (Gharechahi et al. 2014). The differential spots detected on the 2DE gel are digested by trypsin and further analyzed by MS. The digestion of each of the spot of 2DE by proteases gives several peptides and generates a level of complexity which can be easily managed by matrix-assisted laser desorption/ionization (MALDI)-MS/ MS or nano-electrospray liquid chromatography (LC)-MS/MS approaches. This method has been found to be a better approach than other strategies like shotgun method in finding of the novel peptides produced under oxidative stress (Rabilloud et al. 2005). In fact minute amount of proteins present in electrophoretic gel give a MS identification. At any rate, MS/MS approaches often allow both the description and the localization of the modification in the modified peptide. Several heavy metal stress-responsive proteins were identified and characterized in roots of various rice cultivars (*O. sativa*) (Wang et al. 2013a, b; Arenhart et al. 2013, Yang et al. 2013), *Glycine max* (Duressa et al. 2011, Zhen et al. 2007), *Solanum esculentum* (Zhu et al. 2009), *Ananas comosus* (Chen and Lin 2010), *Arabidopsis thaliana* (Karuppanapandian et al. 2012), *Lotus corniculatus* (Navascues et al. 2012), *Phaseolus vulgaris* (Badowiec and Weidner 2014), and *Hordeum vulgare* (Dai et al. 2013) through MALDI-TOF/TOF MS or MS/MS.

Wan and Liu (2008) have investigated the changes in protein expression of rice coleoptiles under oxidative stress triggered by hydrogen peroxide by comparative proteomic approach involving 2DE MALDI TOF MS/MS. Analysis of 2000 spots on each 2DE gel revealed around 150 differentially expressed proteins. Out of them

65 protein spots exhibit upregulation, and 79 were downregulated under oxidative stress. Out of which 129 differentially expressed protein spots were identified by MS to match 89 different protein species having different cellular functions pertaining to maintenance of redox homeostasis, signal transduction, protein metabolism, photoprocesses, and energy metabolism. The similar approach has been used to identify about 43 differentially expressed proteins in wild type and γ-ECS overexpressed transgenic *Nicotiana* in response to drought stress (Kumar et al. 2014). Similarly drought stress-responsive proteins were also identified in cotton using 2DE coupled with MS/MS approach (Zheng et al. 2014).

Proteomic analyses were also been used by Cui et al. (2005) to investigate and understand molecular adaptation mechanisms of chilling stress of rice involving 2DE MALDI TOF MS/MS and ESI MS/MS. Sixty protein spots were found to be upregulated in responding to chilling stress. Out of which, 41 proteins have been characterized by MALDI TOF MS and have been categorized into groups that include factors of protein biosynthesis, molecular chaperones, proteases, enzymes of cell wall biosynthesis, antioxidative defense enzymes, and proteins of energy pathway and signal transduction.

Diagonal gel electrophoresis (DIGE) technique is another one, which is used for proteomic analysis. It is a kind of two-dimensional analysis useful for investigating the subunit composition of proteins containing interchain disulfide bonds or quaternary structure. For this, proteins are firstly electrophoresed under native conditions in the first dimension. After that, they are reduced in the gel and then layered onto a second gel and electrophoresed at 90 °C to the original direction (Samelson 2001). The proteins in the second gel migrate at right angles to their original, first-dimension migration. The proteins which lack in disulfide bond will fall on the diagonal. The spots on the top of the diagonal spot represent the proteins with intramolecular disulfide bridges, whereas upon reduction, the components of the proteins connected by interchain disulfide bonds are resolved below the diagonal line (Samelson 2001). This method is effective to identify stress-responsive modification of disulfide-linked proteins.

COFRADIC (combined fractional diagonal chromatography) method has been developed to study the stress-responsive posttranslational modifications in proteins. In this method, the high complexity of proteome-derived peptide mixtures are reduced, and based on the change in retention time, specific peptides are separated on the same reverse phase column (Gevaert et al. 2006). This procedure has been used for sorting methionyl peptides (Gevaert et al. 2002) as well as cysteinyl peptides (Gevaert et al. 2004). The method involves firstly the reduction and then reaction with DTNB or Ellman's reagent (5,5-dithiobis[2-nitrobenzoic acid]). This reaction leads to the development of heterodisulfide bridge connecting a hydrophobic nitrobenzoic acid group and thiol group of cysteine. The hydrophobic group from the complex gets released immediately after the first chromatographic run. From here, thiol-containing proteins start showing more hydrophilic nature and remains with less retention time. Now the proteins of importance get eluted few minutes before the elution time in the first run and can be identified through LC-MS/MS analysis. Now days, this technique has attracted the attention of researchers

7.2.3 Strategies to Study Posttranslational Modifications in Plants in Response to Stress

As under different kinds of stress responses the data comparison (comparative proteomics) and proteome integration analysis becomes very complex, it provokes the workers to the use of posttranslational modifications such as thiol modification and phosphorylation as an alternative factor to study the stress signaling functions. Several strategies have been developed to identify the posttranslational modifications in the plants under stress. In particular, the development of 2D-PAGE technology coupled with the application of 5′-iodoacetamidofluorescein (5′IAF) (Cuddihy et al. 2008, Baty et al. 2002) or MS and 2D-differential in-gel electrophoresis (DIGE) (Fu et al. 2008) enables the identification of stress-related oxidized and reduced proteins. The comparative behavior of salt-responsive membrane-associated proteins in *Arabidopsis thaliana* and *Thellungiella salsuginea* leaf microsomal proteins using 2D-DIGE approach has been studied (Vera-Estrella et al. 2014). Protein phosphorylation mechanism is considered as an important signaling aspect of environmental stress and is one of the most important posttranslational modifications (PMT) that modulate the activity of proteins, protein-protein interaction, and cellular localization (Khan et al. 2005). The differential expression of phosphoproteins induced by water stress in *O. sativa* was detected using a 2DE coupled with MALDI TOF MS (Ke et al. 2009). Similarly, other proteomic techniques as targeted approach, to identify protein modifications that enable us to identify protein modifications without the need of the gel have been developed.

7.2.3.1 Gel-Free Proteomic Techniques for Elucidations of Posttranslational Modifications (PTMs)

Many gel-free proteomic techniques have been developed in the recent times for differential proteome analysis and identification of posttranslational changes under oxidative stress. Multidimensional protein identification methods (MudPIT) (Koller et al. 2002), isotope-coded affinity tags (ICAT) (Griffin et al. 2001), and isobaric tags for relative and absolute quantification (iTRAQ) deserve special attention in this regard. However, limited studies have been performed on posttranslational modifications in plants due to many limitations like the need of greater amount of protein and changes in protein modification at different stages of growth and development of the plant.

The separation and characterization of highly complex protein mixtures from whole organisms require the use of some alternative potential separation methods (Giddings 1987). Multidimensional protein identification technology (MudPIT) has been developed for such specific purpose of identification of posttranslationally modified proteins (Link et al. 1999, Wolters et al. 2001). It involves the steps like (i) generation of peptides from a complex protein mixture, (ii) separation on a strong cation exchange phase in the first dimension, and (iii) reverse phase chromatography in the second dimension. A popular approach to sample fractionation is to use 1D-PAGE of a protein sample and then tryptic digestion of sample to generate peptides. These peptides were then identified by high-throughput capillary multidimensional liquid chromatography coupled to electrospray ionization ESI MS/MS. Thus, an MS/ MS spectrum can be algorithmically compared with predicated peptide spectra from a sequence database to identify the respective protein. Shotgun methods are being developed not only to identify the proteins in a mixture but also to compare the relative level of protein expression in different biological samples (MacCoss et al. 2002).

Gygi and Aebersold (2000) have dealt for the solution of comparative quantification of complex mixture proteins by employing an innovative, multifunctional tagging reagent, termed as isotope-coded affinity tag (ICAT). In fact, the isotope-coded affinity tags (ICAT) are applied in redox proteomics and originally developed to compare complex protein mixtures produced under oxidative stress (Gygi et al. 1999). A marker consisting of three different parts are used in this technique: (a) a thiol reactive compound (an iodoacetamide analogue), (b) a linker containing either heavy or light isotopes, and (c) a biotin tag for separation by avidin-coupled affinity. In this process only free cysteines are accessible. Sethuraman et al. (2004) challenged the protein samples with hydrogen peroxide and then labeled these samples differentially with the heavy or light form of the ICAT. After that, the samples were mixed together and allowed for tryptic digestion. These digested samples were then separated by affinity chromatography. The captured peptides were analyzed by LC-MS. As oxidized cysteines are not susceptible to labeling with ICAT, therefore the intensity was found to decrease for such samples (Sethuraman et al. 2004).

iTRAQ, one of the best techniques used for the differential proteomic research that involve isotope labeling, is developed for the absolute and relative quantification of proteins (Wang Linqian and Zhiguang 2010). In this approach, protein samples are labeled with different dyes (cyanine) and are subjected to differential in-gel electrophoresis (DIGE). The differentially expressed gel spots are then excised and are analyzed through mass spectrometry involving a combination of multidimensional liquid chromatography and tandem mass spectrometry (Zieske 2006, Wu et al. 2005, Aggarwal et al. 2006). This technique reduces the analytical bias caused by different hydrophilicities, abundances, or isoelectric points, and the applied isotope labeling increases the sensitivity and accuracy as it does not involve typical gel electrophoresis. iTRAQ method has been proved as an effective method for examining the differential expression of proteins. More than three thousand redox-sensitive proteins were identified in wheat seedlings by iTRAQ-based quantitative proteomic analysis. iTRAQ labeling was also used to assess proteome changes in *Vitis vinifera* under heat stress (Liu et al. 2014). Similarly, iTRAQ-based quantitative proteomics approach was also used in rice roots for comparing the expression profiles of proteins in response to Al(3^+) stress (Wang et al. 2014a, b). The study revealed the protein profile of *Arabidopsis* root tissue under iron homeostasis using iTRAQ (Lan et al. 2011).

However, limited studies have been performed on PTMs on plants due to:

(i) The changes of PTMs that occur at different stages of development which make it very hard to identify.
(ii) Problem in applying the technique to plant due to the need of larger amount of protein.
(iii) The changing proteins, which is detected as mass changes, require further processing to transform the data into information.

7.2.3.2 Proteomic Approaches for the Determination of Oxidized Cysteines

Proteomic approaches are widely used for the determination of oxidized cysteines in proteins with the following redox informations:

(i) Identification of specific oxidized cysteine residue
(ii) Characterization of protein containing oxidized cysteines
(iii) Identification of disulfide bond connectivity
(iv) Quantification of relative abundance of oxidized and reduced cysteines

Alkylations of cysteine containing proteins are often done for the detection of oxidized cysteine in plants as the direct detection of the same is not possible. Targeting the specific cysteine in polypeptide by specific reagent treatment or enzyme can identify the modified cysteine as the alkylating reagent cannot always distinguish between oxidized and reduced cysteine. When used in conjugation with 2DE-coupled MS, radiolabeled alkylating agents or fluorophores help us to identification of oxidized cysteines. Comparison of spot intensities before and after the treatment of DTT or Trx is a practice in that case (Alvarez et al. 2009). DIGE is another potential technique used to minimize variations of 2DE result. The redox-based DIGE with two samples with or without oxidative stress are labeled with similar thiol-specific fluorophores that have similar quantum yield but different excitation and emission maxima (Cy3 and Cy5), and subsequently by comparing the fluorescence intensities, the level of cysteine oxidation can be monitored. This type of technique seems to have great potential to identify the cysteine containing thiol oxidation under oxidative stress in photosynthetic tissue.

Alternatively the ICAT technology is exploited to quantify cysteine oxidoreduction (Gygi et al. Gygi et al. 1999). In this process, two samples (oxidatively stressed or control) are alkylated with two ICAT reagents (biotinylated IAM-based reagent labeled with ^{12}C and ^{13}C), followed by affinity chromatography in avidin column. Finally the alkylated proteins are digested and characterized by LC-MS. By comparing the intensities for peptides alkylated with ^{12}C- and ^{13}C-labeled ICAT reagent in MS, the abundance of alkylated cysteine residues in two samples may be compared.

Similarly, the identification of disulfide bond connectivity in proteins can be achieved through MS techniques, with precaution that the disulfide-forming cysteine residues remain bonded during the sample preparation. The process involves the following steps:

(i) The proteins are separated on nonreducing SDS-PAGE
(ii) In-gel chemical reduction
(iii) PAGE separation in orthogonal dimension

The proteins containing no disulfide or intramolecular disulfide bonds will show similar migration rate and exhibit a diagonal band, whereas the protein with intermolecular disulfide bond exhibits lower position (below the horizontal line) in the gel.

7.2.4 Strategies to Study S-Glutathiolation and N-Nitrosylation of Proteins

Glutathionylation of proteins, which is commonly termed as thiolation, has been established as a widespread reversible posttranslational modification of proteins that occurs in cells exposed to oxidative stress (Klatt and Lamas 2000). Several methods are employed for identification of proteins undergoing S-glutathionylation during the oxidative stress (Ghezzi and Bonetto 2003). Earlier metabolic labeling of the GSH pool with radioactive [^{35}S] was extensively used, and now it has been coupled with modern proteomic technology to allow identification of S-glutathionylated proteins (Fratelli et al. 2002). S-Glutathionylation, which alters the charge of a protein, can be detected through Western immunoblotting of samples that has undergone IEF gel electrophoresis (Ghezzi et al. 2001). The IEF gel shifts of S-glutathiolation proteins can be reversed by treatment with reducing agents such as dithiothreitol. However, this method suffers serious shortcomings like the protein has to be separated by IEF, along with the availability of antibody. Some authors have used N-biotinyl analogues of reduced cysteine, GSH, or GSH ethyl ester in the study of protein S-thiolation (Sullivan et al. 2000, Eaton et al. 2002). However, this method can be able to identify the proteins that are present in large quantities. Also, the proteins which are extensively glutathionylated under control conditions could not be identified by this method as they can't incorporate labeled GSH further because of its fully oxidation state.

S-Nitrosylation of the proteins is another kind of posttranslational modification, which is less stable and can't withstand denaturing conditions of SDS-PAGE. In an effort to identify this, a biotin switch method has been discovered for the detection of protein S-nitrosylation after SDS-PAGE and electroblotting (Jaffrey and Snyder 2001). In this case, the free thiols are blocked by incubation with a thiol-specific methylating agent (methyl methanethiosulfonate) in the presence of sodium dodecyl sulfate to ensure access to buried cysteine residues. Methyl methanethiosulfonate does not react with nitrosothiols or disulfide bonds under the reaction conditions used. After that, nitrosothiol groups in the proteins are selectively decomposed to thiols with the addition of ascorbate. Finally the free thiol groups are allowed to react with the sulfhydryl-specific biotinylation reagent, biotin-HPDP. The labeled proteins are then separated by SDS-PAGE, electroblotted, and detected by standard immunoblotting methods using anti-biotin antibodies. Proteins which got labeled can also be purified using streptavidin affinity chromatography as well (Sell et al. 2008). Now the biotinylated proteins will be digested by trypsin, and the resulting peptides containing target cysteines for S-nitrosylation were purified by pull down using NeutrAvidin and identified by MS techniques.

HIS-TAG switch method (Camerini et al. 2007) is used for the detection of S-nitrosylation of proteins. It follows the steps like:

(i) Free thiols in proteins are firstly blocked by N-ethylmaleimide (NEM) alkylation.
(ii) S-nitrosothiols are reduced by ascorbate.
(iii) The detection is done using His-tag alkylation.
(iv) Finally the purification and the unambiguous identification of the modified cysteines were done through mass spectrometry.

7.2.5 Proteomic Approach to Identify Carbonylation of Proteins Under Oxidative Stress

Formation of free carbonyl groups through oxidation of proteins is a common feature of plant under environmental stresses. In this mechanism, the side chain of the amino acids (proline, threonine, lysine, and arginine) gets oxidatively converted (carbonylation) into carbonyl (ketone groups or reactive aldehydes) which subsequently result in inactivation, cross-linking, or breakdown of proteins (Dalle-Donne et al. 2003; Davies 2005). In a comparative study of wheat leaves, carbonylation was found to be more in mitochondria than in chloroplast and peroxisomes, demonstrating higher vulnerability of oxidative damage in mitochondria (Bartoli et al. 2004). The chloroplastic PSII subunits (CP47, D2, and D1) which contain reactive groups that covalently bind amines and phenylhydrazine can be detected using tandem mass spectrometry (Ouellette et al. 1998; Anderson et al. 2004). The oxidation of carbonyl group in the proteins can also be detected by estimating the free carbonyl groups through its reaction with di-nitro phenyl hydrazine (DNPH) and determined spectrophotometrically (Levine et al. 1994). This DNPH-based quantitative and readily adaptable technique has been used to determine oxidative stress during UV-B exposure in *Arabidopsis* and bean leaves (Landry et al. 1995; Shi et al. 2005), heavy metal, and chilling stress in maize (Boscolo et al. 2003; Prasad 1996). This DNPH system is used to measure protein oxidation through immunodetection and immunoenrichment by making the use of commercially available anti-DNP antibodies (Shacter et al. 1994). The modified proteins (carbonylated) can be identified by Western blotting as well (Keller et al. 1993). Immunoblotting using distinct antibodies against different reactive carbonyls revealed the accumulation of RCS-modified proteins in leaves in response to salt stress treatment. Coupling the anti-DNP antibody with 2DE can generate more authentic data regarding protein carbonylation (Job et al. 2005). For this, the proteins were initially separated 2DE, and then the oxidized proteins were tagged by Western blotting. The difficulty in solubilizing membrane proteins for the first dimension, limited and the poor resolution of some of the protein groups, has compelled the workers to use immunopurification methods for enriching the DNP-derivatized proteins (Kristensen et al. 2004;

Davletova et al. 2005a). In that case, these artifacts can be removed by using the antibody-based column purification before the gel electrophoresis or even by purifying the proteins through immunoaffinity columns. Finally the proteins can be analyzed through LC-MS/MS.

The major finding reported for oxidative modifications protein under in rice, especially those that are sensitive to different stresses, exhibits identification of large number of oxidatively modified proteins including posttranslational modification which play vital role in redox regulation (Ali and Komatsu 2006; Abbasi and Komatsu 2004; Parker et al. 2006). In addition to the vital role antioxidantive defense as redox regulatory proteins that are expected to be identified in stressed plant, another protein group that functions in energy metabolism stands out as it seems to be activated almost in every stress. The posttranslational modification (PTM) such as phosphorylation and thiol modification caused by oxidative stress could be an alternative study to investigate stress signaling functions. Several strategies have been developed to identify PTM in rice. In particular, the development of 2D-PAGE technology coupled with the application of 5′-iodoacetamidofluorescein (5′IAF) or MS such as in 2D-fluorescense difference gel electrophoresis (DIGE) enables the identification of stress-related oxidized and reduced proteins in rice (). Many gel-free proteomic systems have also been developed for differential proteome analysis (already descried). Examples are multidimensional protein identification methods (MudPIT) that effectively identify individual components of proteins or peptides by avoiding band broadening for chromatographic detection; isotope-coded affinity tags (ICAT) and isobaric tags for relative and absolute quantification (iTRAQ) are also conventionally used for the detection of redox regulatory proteins. These techniques are considered a targeted method to identify modification in proteins by means of the mass differences between oxidative stressed and unstressed proteomes.

7.3 Genomic Approach for Understanding Oxidative Stress Responses in Plants

Large-scale characterization of redox regulatory genes that help plants to tolerate oxidative stress is essential to understand genetics of plant stress adaptation and to successfully develop transgenic plants to be grown under adverse conditions. In fact, the initial strategy for improving crop production involves identification of redox-regulatory genes. Nevertheless, association of the functions of different genes with certain biological processes can also be achieved by identifying the expression patterns of these genes, especially under stress conditions (Huang et al. 2014). Moreover, oxidative stress in plants also leaves traces of biomarkers at genomic level, which could be identified with technological breakthrough involving the approaches of genomics and transcriptomics followed by incorporating the data in bioinformatics for validating the same.

7.3.1 Transcriptomic Study

The major focus of transcriptome profiling and/or transcriptomics includes (a) identification of genes involved in oxidative stress tolerance, (b) gene expression quantification, and (c) unraveling mechanisms underlying the response of the gene expression plasticity to stress factors (Schulze and Downward 2001; Stears et al. 2003).

7.3.1.1 Transcript Profiling

Understanding differential expression of gene is the basis for any kind of study pertaining to plant – environmental stress interaction – where the identification and measurement of the specific transcript (mRNA) sequences may be the sensitive parameter to unfold plant responses (Nicot et al. 2005). The following sensitive analytical methods were developed to perform high-throughput profiling of the gene space (Liang and Pardee 1992; Velculescu et al. 1995; Schena et al. 1995; Bachem et al. 1996; Brenner et al. 2000; Busch 2007):

(i) Serial analysis of gene expression (SAGE)
(ii) Microarray
(iii) cDNA-amplified fragment length polymorphism (AFLP)
(iv) Massively parallel signature sequencing (MPSS) and the ATH1 Genome Array

Hybridization-based microarrays have been developed to analyze transcriptional changes in many plant species. Microarray-based transcriptomic studies have been evidenced in the characterization of transcriptomes in a number of plant species (Strable et al. 2008; Auge et al. 2009). Microarray data on stably expressed genes has been successfully employed in several plants including *Arabidopsis* (Czechowski et al. 2005; Dekkers et al. 2012), rice (Narsai et al. 2010), etc. A typical microarray platforms have been reported to exhibit a limited dynamic range of several orders of magnitude (Schena et al. 1995; Wang et al. 2009) because of the restriction in microarray dynamic range due to the density/availability of probe, fluorescent dye the intensity, and the sensitivity of scanning instruments (reviewed by Wang et al. 2010).

The utility of the microarray system for identifying H_2O_2-responsive genes may be illustrated by the demonstration that ESTs corresponding to the antioxidant genes like *PAL1* and *GST6*. It was found that H_2O_2 responsiveness in *Arabidopsis* suspension cultures (Desikan et al. 2000) was among those identified via microarray. Expression data and sequence homologies (based on EST description and BLAST homologies) for H_2O_2-responsive ESTs exhibited induction greater than fivefold, including those encoding a heat shock protein, a zinc finger protein, calmodulin, and an ethylene-responsive transcription. Further ESTs with transcripts of low abundance (i.e., those with low channel intensities post-hybridization) that were H_2O_2 inducible were also identified. These included ESTs encoding potential signaling proteins such as a heat shock transcription factor, a His kinase, and a protein Tyr phosphatase.

7.3.2 Genome-Wide Association Studies (GWAS) and Quantitative Real-Time Reverse-Transcriptase PCR Approach for Understanding Oxidative Stress Responses in Plants

Plant responses to environmental stress are mainly polygenic in nature. Identification of genes responsible for intraspecific variation in plant abiotic stress response is therefore a prerequisite (Verslues et al. 2014). Genome-wide association (GWAS) and quantitative real-time reverse-transcriptase PCR (qRT-PCR) are the prime approaches to identify the genes responsible for stress tolerance. Through GWAS, it is now possible to identify critical inducible genes and their alleles apart from successfully identifying loci for phenotypic variation (Ma et al. 2012). For unfolding the potential association between genetic variation and important agricultural traits, this method was widely exploited (Brachi et al. 2011; Ma et al. 2012). GWAS have been found to map gene with high resolution. GWAS with the single-nucleotide polymorphism (SNP) typing can identify small haplotype blocks. That ascorbate peroxidases are member of a multigenic family has been confirmed by GWAS (Najami et al. 2008). GWAS also helped to identify genomic regions that were correlated with Al tolerance (Famoso et al. 2011). Following GWAS and reverse genetics approach, dehydration stress-induced proline accumulation in *Arabidopsis thaliana* was identified (Verslues et al. 2014). In molecular genetics, using *A. thaliana* and *O. sativa*, the role of GWAS for identifying the genetic bases of adaptive traits was validated (Brachi et al. 2011). The potential of GWAS in *A. thaliana* by successful functional validation of the gene *ACCELERATED CELL DEATH6* (*ACD6*) has been established. While ACD6 has been formerly identified from forward-genetic mutant screens, the study established the capability of GWAS in the detection of true positives (Lu et al. 2003). In spite of the fact that environmental stresses significantly impact Soybean max production, GWAS has been used rarely in this plant.

Quantitative real-time reverse-transcriptase PCR (qRT-PCR) was proposed to sort out the major drawbacks of typical microarray platforms, where analysis of differences in gene expression is possible. qRT-PCR is known for its sensitivity, accuracy, and specificity (Vanguilder et al. 2008). It can also able to quantify transcript levels of genes with fairly low transcript levels (Nolan et al. 2006). However, to ensure accurate qRT-PCR assessment of expression in genes of interest, it was considered necessary to maintain the stability of the expression of reference genes and data reliability (Wang et al. 2014a, b). In view of the fact that several internal control genes that are referred in qRT-PCR exhibit fluctuation during treatments, it is therefore recommended to avoid the bias (Nicot et al. 2005). Another fact is that only a snapshot of information is constituted by qRT-PCR regarding the quantity of a given transcript in a cell or tissue. So it may be recommended to include additional information regarding regulatory RNAs, protein levels, and protein activity to assess the biological consequences of variable mRNA levels.

Both qRT-PCR and semi qRT-PCR methodologies have been developed for the quantification of gene expression in a number of plants (rice, sugar beet, etc.) under conditions of environmental stress (Chakraborty and Bhattacharjee 2015; Taski-Ajdukovic et al. 2012). Due to the presence of distinct transcriptomes in seeds compared with the vegetative tissues, revalidation of reference genes in samples of seed development and germination has been considered necessary (Li et al. 2012).

7.4 Genetic Engineering for Oxidative Stress Tolerance

Considering a significant loss in crop productivity due to abiotic stresses, the development of transgenic plants with desirable traits has been the focus of current plant research. The prime initial approach to improve crop production involves transformation of useful and functional antioxidant genes into crop through an effort of genetic engineering to produce stress-tolerant crops (Table 7.1), though it was found that common antioxidants are not appropriate for all types of stress and hence stress-specific traits are needed to effectively targeted the different stress signaling pathways. Global analysis of transcription levels in rice plants under different abiotic stresses and in different organs has been intensively studied; a few examples include high salinity (Vera-Estrella et al. 2014), drought (Khan et al. 2005), high temperature (Ke et al. 2009), cold and H2O2 (Koller et al. 2002), and light (Griffin et al. 2001). However, in natural conditions plant is exposed to various stresses simultaneously, and, therefore, the responses toward stress adaptation are expected to be different compared to those exposed to single stress condition. Moreover, since in most of the cases, abiotic stresses work via oxidative stress, tracking the expression of redox regulatory genes may be the target. Several workers have proved the existence of significant cross talks between stress signaling pathways through the central player ROS on gene transcription integration. In fact, the complex stress regulatory system is controlled by ABA-dependent or ABA-independent patterns. Dehydration-responsive element binding protein (DREB2) and NAC are examples of ABA-independent TFs (Link et al. 1999; Wolters et al. 2001), whereas the ABA-responsive element (ABRE) binding protein factor (ABF) functions in ABA-dependent gene expression (MacCoss et al. 2002) although in both the cases, ROS play a vital role either by integrating through feed forward or back work processes. In rice, several DREB cDNAs were characterized, namely, OsDREB1A, OsDREB1B, OsDREB1C, OsDREB1D, and OsDREB2A (Sethuraman et al. 2004); OsDREB1A and OsDREB1B were found to be induced by cold stress, while OsDREB2A was found to have the ability to regulate salt and drought stress (Sethuraman et al. 2004). Overexpression of OsDREB1A in rice could induce expression of DRE/CRT genes. It has been reported that the OsDREB1A overexpression lines show phenotypes similar to AtDREB1A (*Arabidopsis*) overexpression line, but with improved stress tolerance involving through redox regulation (Sethuraman et al. 2004).

Table 7.1 Examples of some transgenesis for developing oxidative stress-resistant crops

Target gene	Source of gene	Transgenic developed	Response	Reference
Mn SOD	Yeast	Rice transgenic plants maintained high level of SOD	Plant maintained high level of SOD and ascorbate peroxidase activity in chloroplast upon exposure to salt stress, while decrease in SOD activities was observed in wild-type plants	Tanaka et al. (1999)
Mn SOD	*Pisum sativum*	Rice transgenic plants resistant to dehydration stress	Better membrane permeability status and integrity of transgenic plants compared to that observed in wild-type plants under dehydration stress imposed by PEG	Wang et al. (2005)
Cu/Zn SOD	*A. marina*	Rice transgenic plants resistant to oxidative stress	Plants more tolerant to methyl viologen (MV)-mediated oxidative stress	Prasanth et al. (Prashanth et al. 2008)
OsAPXa	Rice	Rice transgenic plants resistant to chilling stress	Plants showed reduced lipid peroxidation and oxidative damage when subjected to cold stress	Sato et al. (2011)
OsAPxl and OsAPx2	Rice	Rice transgenic plants exhibiting better redox regulation	Compensation of ascorbate peroxidase by other peroxidases required for redox homeostasis	Bonifacio et al. (2011)
CAT	*Triticum aestivum* L.	Rice transgenic plants resistant to chilling stress	Redox homeostasis mentioned by up-regulation of CAT activities in transgenic plants compared to that observed in wild-type plants	Matsumura et al. (2002)
CAT	*Suaeda salsa*	Rice transgenics resistant to salt- and paraquat-induced oxidative stress	Redox homeostasis mentioned by up-regulation of GST and reduction of oxidative membrane damage	Zhao and Zhang (2006)
APx1/2s	Rice	Rice transgenics resistant to salt- and temperature-induced oxidative stress	Up-regulation of other peroxidases was observed under salinity, heat, high light, and MV treatment	Bonifacio et al. (2011)
GST	Rice	Rice transgenics resistant to submergence and chilling-induced oxidative stress	Transgenic lines exhibiting submergence and chilling stress	Takesawa et al. (2002)

(continued)

Table 7.1 (continued)

Target gene	Source of gene	Transgenic developed	Response	Reference
GST	*Suaeda salsa*	Rice transgenics resistant to salt- and paraquat-induced oxidative stress	Salt and paraquat stress tolerance were observed due to up-regulation in GST, CAT, and SOD activity	Zhao and Zhang (2006)
APOX1	*Arabidopsis* mutant	*Arabidopsis* resistant to heat tolerance	Heat tolerance with better reactive oxygen metabolism	Davletova et al. 2005a

7.4.1 Transgenesis for Oxidative Stress-Tolerant Plants Targeting Antioxidant Genes

The synchronous tandem action of enzymatic (SOD; CAT; GPX; GST; APX; MDHAR; DHAR; GR) and nonenzymatic (ascorbate, glutathione, carotenoids; tocopherols; phenolics) antioxidants regulates the redox status to get rid of consequences potentially caused by ROS and its secondary reaction products (Gill and Tuteja 2010; Anjum et al. 2010, 2012; Gill et al. 2013). Abiotic stress tolerance has been achieved in a number of plants through transgenic approach via engineering ROS scavenging antioxidant (enzymatic and nonenzymatic) components including CAT, APX, SOD, GPX, GR, MDHAR, DHAR, GSH, and AsA (Zhu et al. 1999; Roxas et al. 1997; Chen et al. 2003; Kwon et al. 2002, 2003; Chen and Gallie 2005; Del Rio et al. 2006; Lee et al. 2007; Yin et al. 2010).

Transgenesis of wheat (*Triticum aestivum*) DHAR gene in tobacco (*Nicotiana tabacum*) and maize (*Zea mays*) was reported to enhance DHAR enzyme activity by many times in transgenic *N. tabacum* and *Z. mays* (Chen et al. 2003). The authors noted two to fourfold increased AsA content in *Z. mays* leaves and grains due to overexpression of DHAR. In another study, larger quantities of AsA were evidenced in transgenic *N. tabacum* plants expressing the gene for *T. aestivum* DHAR (Chen and Gallie 2005). Being a significant component of AsA-GSH pathway, GR confers abiotic stress tolerance in plants through its involvement in the detoxification of ROS and regeneration of GSH (Gill and Tuteja 2010; Anjum et al. 2010, 2012; Gill et al. 2013). The role of GR transgenes in conferring abiotic stress tolerance has been credibly reported (reviewed by Gill et al. 2013).

The strategy to improve the rice production involving the transformation of useful antioxidant and redox regulatory genes has been tabulated in Table 7.1.

7.4.2 Transgenesis for Oxidative Stress-Tolerant Plants Targeting Sulfur Metabolism Pathway Genes

Sulfur metabolism has been implicated as a versatile platform for redox regulation in plants (Rausch and Wachter 2005). The list of functionally important S-containing compounds involving in redox regulation includes the tripeptide glutathione (GSH), cysteine (Cys), methionine (Met), thioredoxins (Trx), and peroxiredoxin (Prx) (Saito 2000; Khan et al. 2008; Gill and Tuteja 2011). Nevertheless, owing to the significance of sulfur and sulfur-containing products in plants, a number of plants have been constructed with overexpressing thiol and sulfur-metabolizing enzymes and/or products including glutathione reductase, glutathione S-transferase, cysteine synthase (CSase), or serine acetyltransferase (SATase) gene, key enzymes for cysteine (Cys) biosynthesis in plants. Moreover, the overaccumulation of sulfur-containing compounds, Cys and GSH, has also been reported in transgenic plants overexpressing SATase gene (Noji and Saito 2007). Reports are available on the transgenic plants increasing the level of sulfur-containing metabolites (Wirtz and Hell 2003) and tolerant to oxidative stress (Youssefian et al. 2001). Transgenic *N. tabacum* plants overexpressing CSase exhibited enhanced Cd tolerance (Kawashima et al. 2004). Activation of Met4 gene and subsequent upregulation of tripeptide GSH was reported (Barbey et al. 2005). Cd-mediated upregulation of Sultr1;1 and Sultr2;1 genes encoding two sulfate transporters has also been found after 2 or 6 h of Cd treatment and 12–24 h after sulfate depletion (Takahashi et al. 2000; Herbette et al. 2006; Sarry et al. 2006).

7.4.3 Transgenesis for Oxidative Stress-Tolerant Plants Targeting Transcription Factors and Signal Transduction Pathways

Proteins involved in differential expression of genes (promoter and repressor) are designated as transcription factors (TFs). For evoking such activity, i.e., to promote or obstruct RNA polymerases to the DNA template, TFs act together with other transcriptional regulators, like chromatin remodeling/modifying proteins (Udvardi et al. 2007). TFs have been reported to mediate plant stress acclimation or adaptation through interacting with cis-elements in the promoter regions of several stress-related genes (Agarwal and Jha 2010). As stress tolerance can be controlled by TFs both in homologous and heterologous expression systems, TFs have been the focus of genetic manipulation to raise stress-tolerant plants (Pitzschke 2012, 2013). Moreover, the manipulation of complex metabolic pathways in plants is being accomplished by employing TFs because these tend to regulate and manage multiple signaling and metabolic pathways in plants, particularly under stress (Hussain et al. 2011).

In plants, a number of processes for stress adaptation including rapid signal perception or sensing and adaptation to suboptimal conditions are controlled by a network of robust and densely interconnected signaling systems with several cross talk (Danquah et al. 2014; Smékalová et al. 2014). The ABA (abscisic acid) and MAPK (mitogen-activated protein kinase) signaling pathways are considered as a central regulator of abiotic stress-induced oxidative stress response in plants since these signaling pathways trigger major changes in gene expression and plant stress-adaptive physiological responses (Seki et al. 2002; Yamaguchi-Shinozaki and

Shinozaki 2006; Ning et al. 2010). Recent studies reported that ABA-perception core signaling complexes can transmit cues to downstream events (Fujii et al. 2009; Ma et al. 2009; Park et al. 2009). The position of ROS in ABA signaling pathway seems both feed forward or backward. To the other, MAPK signaling pathways are sophisticated multi-enzymatic complexes and are involved in plant stress-adaptive and acclimation processes (Colcombet and Hirt 2008; Rodriguez et al. 2010; Sinha et al. 2011). The role of MAPKs in abiotic stress-induced redox regulation and signaling in plants has been extensively reported. Under extremes of temperature, MAPK activation induces physiological response-mediated changes in gene expression (Huang et al. 2012; Mishra et al. 2006). The role of MPK4 and MPK6 for the cold stress responses has also been reported in *Arabidopsis* (Teige et al. 2004).

Participation of ROS or redox status, in cell signaling, homeostasis, and MAPK signaling, has also been reported (Smékalová et al. 2014). Elevated ROS level-mediated activation of MPK3 and MPK6 in *Arabidopsis* has been noticed (Lumbreras et al. 2010). Oxidative stress-accrued activation of MPK3/MPK6 activation was reported in *Arabidopsis*, where oxidative signal-inducible 1 gene (OXI1) was considered necessary for oxidative burst-mediated signaling (Rentel et al. 2004). Activation of MPK1 and MPK2 (Ortiz-Masia et al. 2007), MPK3 and MPK6 (Kovtun et al. 2000), MPK4 (Nakagami et al. 2006), and MPK7 (Doczi et al. 2007) due to oxidative stress imposed by exogenous application of H_2O_2 was evidenced in *Arabidopsis*. It has been reported that MAPK phosphatase 2 (MKP2) can interact with MPK3 and MPK6 in *Arabidopsis* plants, and its activation can be triggered by loss of redox homeostasis. Since MKP2 dephosphorylates MPK3 and MPK6, it promotes oxidative stress tolerance in plants (Lumbreras et al. 2010). There are evidences of the transmission of phospho-signals by MKK5 to both MPK6 and MPK3 that in turn mediated plant responses to ozone-related oxidative stress (Miles et al. 2009).

That the over expression of TaASR1 found to induce tolerance to oxidative stress could be vouched by lesser lipid peroxidation, ion leakage (IL), and ROS accumulation (Hu et al. 2013). In *Arabidopsis*, abiotic multiple stress tolerance by overexpressing transcription factor (ABF2), ABA-responsive element (ABRE)-binding factor (ABF) was reported (Kim et al. 2004). Expression of zinc finger protein Zat12 in *Arabidopsis* increased the expression of redox-responsive genes (Davletova et al. 2005b). Overexpression of a bZIP transcription factor gene (PtrABF) was reported to enhance dehydration and drought tolerance in tobacco via redox regulation and modulating expression of redox-sensitive antioxidative genes (Huang et al. 2010). The modulation of peroxidase-mediated scavenging of H_2O_2 by *Poncirus trifoliata*-PtrbHLH, a basic helix-loop-helix TF in *N. tabacum* and lemon, was reported (Huang et al. 2013). In *T. aestivum*, the overexpression of TaAQP7 found to increase tolerance to drought-mediated oxidative stress in *N. tabacum*. The transgenic *N. tabacum* lines exhibited lower levels of lipid peroxidation and ROS (Zhou et al. 2012). Similarly, the overexpression of WRKY genes in *Arabidopsis* was evidenced to provide oxidative stress tolerance and subsequent protection (Davletova et al. 2005a; Gadjev et al. 2006). The transgenic *N. tabacum* overexpressing WRKY10 resulted in enhanced drought and salt stress tolerance by lower ROS and lipid peroxidation (Wang et al. 2013a, b).

7.5 Conclusions and Future Perspectives

Redox Omic Science, developed in last couple of decades, is increasingly gaining importance to unfold plant responses to environmental stress. Proteomics has proved to be a precious tool to identify and characterize proteins involved in plants responses to environmental stress. Last few years also witnessed tremendous development in identifying large number of plant proteins that are subjected to redox modifications. The identification of redox PTMs of proteome developed under abiotic stress-induced oxidative threat deserves special attention in this regard. The mechanism of ROS sensing through thiol modifications of target protein was unfolded in last decade and helps us to understand the preliminary event of ROS signaling. However, deciphering the redox PTMs on plant proteome level is still far from complete due to many technical reasons, like the artificial oxidations of tryptophan, methionine during sample preparation, larger complexity of redox regulatory protein families, apoplastic and chloroplastic contribution in redox regulation, etc. Apart from that, mass spectrometry is still not sufficiently a sensitive technique to characterize, quantify, and localize the posttranslational modifications, when it comes to the level of change in a small number of amino acid residues within the protein. Thus, quantification of the extent of modification within a protein remains a challenge. Therefore, it is pertinent to improve the technologies for characterizing the PTMs by taking care of all these constraints. Comparative proteomic approach used to compare the expression of redox regulatory proteome in different tissues, organs, and subcellular compartments in time course manner and can contribute significantly in understanding of the mechanism of plant responses to unfavorable environmental cues. Further advancement of high-throughput technologies are required to solve limitations associated with the analytical variability of the technique which can also ensure reproducibility of experimental results. Another *omic* approaches like genomics, transcriptomics, and metabolomics can further help to establish redox regulatory networks of interaction between genes, proteins, and metabolites involved in stress response mechanisms. In this regard, the studies on the characterization of genes underlying antioxidative defense, signaling pathways, transcription factors, and metabolisms should be intensified not only for better understanding of redox regulatory phenomenon under stress but also to successfully develop stress-resistant plants.

References

Abbasi FM, Komatsu S (2004) A proteomic approach to analyze salt-responsive proteins in rice leaf sheath. Proteomics 4(7):2072–2081

Agarwal PK, Jha B (2010) Transcription factors in plants and ABA dependent and independent abiotic stress signaling. Biol Plant 54:201–212

Aggarwal K, Choe LH, Lee KH (2006) Shotgun proteomics using the iTRAQ isobaric tags. Brief Funct Genomic Proteomic 5:112–120

Ali GM, Komatsu S (2006) Proteomic analysis of rice leaf sheath during drought stress. J Proteome Res 5(2):396–403

Alvarez S, Berla BM, Sheffield J, Cahoon RE, Jez JM, Hicks LM (2009) Comprehensive analysis of the Brassica juncea root proteome in response tocadmium exposure by complementary proteomic approaches. Proteomics 9:2419–2431

Anderson LB, Ouellette AJ, Eaton-Rye J, Maderia M, MacCoss MJ, Yates JR, Barry BA (2004) Evidence for a post-translational modification, aspartyl aldehyde, in a photosynthetic membrane protein. J Am Chem Soc 126:8399–8405

Anjum NA, Umar S, Chan MT (2010) Ascorbate-glutathione pathway and stress tolerance in plants. Springer, Dordrecht

Anjum NA, Ahmad I, Mohmood I, Pacheco M, Duarte AC, Pereira E, Umar S, Ahmad A, Khan NA, Iqbal B, Prasad MNV (2012) Modulation of glutathione and its related enzymes in plants' responses to toxic metals and metalloids – a review. Environ Exp Bot 75:307–324

Arenhart RA, Lima JC, Pedron M, Carvalho FE (2013) Involvement of ASR genes in aluminium tolerance mechanisms in rice. Plant Cell Environ 36:52–67

Auge GA, Perelman S, Crocco CD, Sanchez RA, Botto JF (2009) Gene expression analysis of light-modulated germination in tomato seeds. New Phytol 183:301–314

Bachem CW, van der Hoeven RS, de Bruijn SM, Vreugdenhil D, Zabeau M, Visser RG (1996) Visualization of differential gene expression using a novel method of RNA fingerprinting based on AFLP, analysis of gene expression during potato tuber development. Plant J 9:745–753

Badowiec A, Weidner S (2014) Proteomic changes in the roots of germinating *Phaseolus vulgaris* seeds in response to chilling stress and post-stress recovery. Biochem Biophys Acta 1844:1208–1218

Barbey R, Baudouin-Cornu P, Lee TA, Rouillon A, Zarzov P, Tyers M, Thomas D (2005) Inducible dissociation of SCF(Met30) ubiquitin ligase mediates a rapid transcriptional response to cadmium. EMBO J 24:521–532

Bartoli CG, Gomez F, Martinez DE, Guiamet JJ (2004) Mitochondria are the main target for oxidative damage in leaves of wheat (*Triticum aestivum* L.). J Exp Bot 55:1663–1669

Baty JW, Hampton MB, Winterbourn CC (2002) Detection of oxidant sensitive thiol proteins by fluorescence labeling and two-dimensional electrophoresis. Proteomics 2:1261–1266

Bonifacio A, Martins MO, Ribeiro CW, Fontenele AV, Carvalho FEL, Margis-Pinheiro M, Silveira JAG (2011) Role of peroxidases in the compensation of cytosolic ascorbate peroxidase knockdown in rice plants under abiotic stress. Plant Cell Environ 34:1705–1722

Boscolo PR, Menossi M, Jorge RA (2003) Aluminum-induced oxidative stress in maize. Phytochemistry 62:181–189

Brachi B, Morris GP, Borevitz JO (2011) Genome-wide association studies in plants, the missing heritability is in the field. Genome Biol 12:232

Brenner S, Johnson M, Bridgham J, Golda G, Lloyd DH, Johnson D, Luo S, McCurdy S, Foy M, Ewan M, Roth R, George D, Eletr S, Albrecht G, Vermass E, Williams SR, Moon K, Burcham T, Pallas M, DuBridge RB, Kirchner J, Fearon K, Mao J, Corcoran K (2000) Gene expression analysis by massively parallel signature sequencing (MPSS) on microbead arrays. Nat Biotechnol 18:630–634

Busch W, Lohmann JU (2007) Profiling a plant, expression analysis in *Arabidopsis*. Curr Opin Plant Biol 10:136–141

Buss H, Chan TP, Sluis KB, Domigan NM, Winterbourn CC (1997) Protein carbonyl measurement by a sensitive ELISA method. Free Radic Biol Med 23:361–366

Caddihy SL, Baty JW, Brown KK, Winterbourn CC, Hampton MB (2008) Proteomic detection of oxidized and reduced thiol proteins in cultured cells. Methods Mol Biol 519:363–375

Camerini S, Polci ML, Restuccia U, Usuelli V, Malgaroli A, Bachi A (2007) A novel approach to identify proteins modified by nitric oxide, the HIS-TAG switch method. J Proteome Res 6:3224–32331

Chakraborty A, Bhattacharjee S (2015) Differential competence of redox-regulatory mechanism under extremes of temperature determines growth performances and cross tolerance in two indica rice cultivars. J Plant Physiol 176:65–77

Chen Z, Gallie DR (2005) Increasing tolerance to ozone by elevating foliar ascorbic acid confers greater protection against ozone than increasing avoidance. Plant Physiol 138:1673–1689

Chen JH, Lin YH (2010) Effect of aluminum on variations in the proteins in pineapple roots. Soil Sci Plant Nutr 56:438–444

Chen Z, Young TE, Ling J, Chang SC, Gallie DR (2003) Increasing vitamin C content of plants through enhanced ascorbate recycling. Proc Natl Acad Sci U S A 100:3525–3530

Claiborne A, Miller H, Parsonage D, Ross RP (1993) Protein-sulfenic acid stabilization and function in enzyme catalysis and gene regulation. FASEB J 7:1483–1490

Colcombet J, Hirt H (2008) Arabidopsis MAPKs, a complex signalling network involved in multiple biological processes. Biochem J 413:217–226

Cui S, Huang F, Wang J, Ma X, Cheng Y, Liu J (2005) A proteomic analysis of cold stress responses in rice seedlings. Proteomics 5(12):3162–3172

Cumming RC (2008) Analysis of global and specific changes in the disulfide proteome using redox two-dimensional polyacrylamide gel electrophoresis. Methods Mol Biol 476:165–179

Czechowski T, Stitt M, Altmann T, Udvardi MK, Scheible WR (2005) Genome-wide identification and testing of superior reference genes for transcript normalization in Arabidopsis. Plant Physiol 139:5–17

Dai H, Cao F, Chen X, Zhang M (2013) Comparative proteomic analysis of aluminum tolerance in tibetan wild and cultivated barleys. PLoS One 8:e63428

Dalle-Donne I, Rossi R, Giustarini D, Milzani A, Colombo R (2003) Protein carbonyl groups as biomarkers of oxidative stress. Clinic Chim Acta 329:23–38

Danquah A, de Zelicourt A, Colcombet J, Hirt H (2014) The role of ABA and MAPK signaling pathways in plant abiotic stress responses. Biotechnol Adv 32:40–52

Davies MJ (2005) The oxidative environment and protein damage. Biochim Biophys Acta 1703:93–109

Davletova S, Rizhsky L, Liang H, Shengqiang Z, Oliver DJ, Coutu J, Shulaev V, Schlauch K, Mittler R (2005a) Cytosolic ascorbate peroxidase 1 is a central component of the reactive oxygen gene network of *Arabidopsis*. Plant Cell 17:268–281

Davletova S, Schlauch K, Coutu J, Mittler R (2005b) The zinc-finger protein Zat12 plays a central role in reactive oxygen and abiotic stress signaling in Arabidopsis. Plant Physiol 139:847–856

Dekkers BJ, Willems L, Bassel GW, van Bolderen-Veldkamp RP, Ligterink W, Hilhorst HW, Bentsink L (2012) Identification of reference genes for RT-qPCR expression analysis in Arabidopsis and tomato seeds. Plant Cell Physiol 53:28–37

Del Rio LA, Sandalio LM, Corpas FJ, Palma JM, Barroso JB (2006) Reactive oxygen species and reactive nitrogen species in peroxisomes. Production, scavenging and role in cell signaling. Plant Physiol 141:330–335

Desikan R, Neill SJ, Hancock JT (2000) Hydrogen peroxide-inducedgene expression in Arabidopsis thaliana. Free Radic Biol Med 28:773–778

Doczi R, Brader G, Pettko-Szandtner A, Rajh I, Djamei A, Pitzschke A, Teige M, Hirt H (2007) The Arabidopsis mitogen-activated protein kinase kinase MKK3 is upstream of group C mitogen-activated protein kinases and participates in pathogen signaling. Plant Cell 19:3266–3279

Duressa D, Soliman K, Taylor R, Senwo Z (2011) Proteomic analysis of soybean roots under aluminum stress. Int J Plant Genom 2011:1–12

Eaton P, Fuller W, Shattock MJ (2002) S-Thiolation of HSP27 regulates its multimeric aggregate size independently of phosphorylation. J Biol Chem 277:21189–21196

Famoso AN, Zhao K, Clark RT, Tung C-W, Wright MH, Bustamante C, Kochian LV, McCouch SR (2011) Genetic architecture of aluminum tolerance in rice (*Oryza sativa*) determined through genome-wide association analysis and QTL mapping. PLoS Genet 7:e1002221

Foyer CH, Noctor G (2005) Oxidant and antioxidant signalling in plants, a re-evaluation of the concept of oxidative stress in a physiological context. Plant Cell Environ 28:1056–1071

Fratelli M, Demol H, Puype M (2002) Identification by redox proteomics of glutathionylated proteins in oxidatively stressed human T lymphocytes. Proc Natl Acad Sci U S A 99:3505–3510

Fu C, Hu J, Liu T, Ago T, Sadoshima J, Li H (2008) Quantitative analysis of redox-sensitive proteome with DIGE and ICAT. J Proteome Res 7:3789–3802

Fujii H, Chinnusamy V, Rodrigues A, Rubio S, Antoni R, Park SY, Cutler SR, Sheen J, Rodriguez PL, Zhu JK (2009) In vitro reconstitution of an abscisic acid signalling pathway. Nature 462:660–664

Fukuda M, Islam N, Woo S-H, Yamagishi A, Takaoka M, Hirano H (2003) Assessing matrix assisted laser desorption/ ionization-time of flight-mass spectrometry as a means of rapid embryo protein identification in rice. Electrophoresis 24:1319–1329

Gadjev I, Vanderauwera S, Gechev TS, Laloi C, Minkov IN, Shulaev V, Apel K, Inze D, Mittler R, Van BF (2006) Transcriptomic footprints disclose specificity of reactive oxygen species signaling in Arabidopsis. Plant Physiol 141:436–445

Gevaert K, Van Damme J, Goethals M, Thomas GR, Hoorelbeke B, Demol H, Martens L, Puype M, Staes A, Vandekerckhove J (2002) Chromatographic isolation of methionine-containing peptides for gel-free proteome analysis, identification of more than 800 Escherichia coli proteins. Mol Cel Proteomics 1:896–903

Gevaert K, Ghesquière B, Staes A, Martens L, Van Damme J, Thomas GR, Vandekerckhove J (2004) Reversible labeling of cysteine-containing peptides allows their specific chromatographic isolation for non-gel proteome studies. Proteomics 4:897–908

Gevaert K, Van Damme P, Ghesquière B, Vandekerckhove J (2006) Protein processing and other modifications analysed by diagonal peptide chromatography. Biochim Biophys Acta 1764:1801–1810

Gharechahi J, Alizadeh H, Reza Naghavi M, Sharifi G (2014) A proteomic analysis to identify cold acclimation associated proteins in wild wheat (*Triticum urartu* L.). Mol Biol Rep 41:3897–3905

Ghezzi P, Bonetto V (2003) Redox proteomics, identification of oxidatively modified proteins. Proteomics 3:1145–1153

Ghezzi P, Romines B, Fratelli M, Eberini I, Gianazza E, Casagrande S, Laragione T, Mengozzi M, Herzenberg LA, Herzenberg LA (2001) Protein glutathionylation, coupling and uncoupling of glutathione to protein thiol groups in lymphocytes under oxidative stress and HIV infection. Mol Immunol 38:773–780

Giddings JC (1987) Concepts and comparisons in multidimensional separation. J High Resolut Chromatogr 10:319–323

Gill SS, Tuteja N (2010) Reactive oxygen species and antioxidant machinery in abiotic stress tolerance in crop plants. Plant Physiol Biochem 48:909–930

Gill SS, Tuteja N (2011) Cadmium stress tolerance in crop plants – probing the role of sulfur. Plant Signal Behav 6:215–222

Gill SS, Anjum NA, Hasanuzzaman M, Gill R, Trivedi DK, Ahmad I, Pereira E, Tuteja N (2013) Glutathione and glutathione reductase, a boon in disguise for plant abiotic stress defense operations. Plant Physiol Biochem 70:204–212

Gitlin G, Tsarbopoulos A, Patel ST, Sydor W, Pramanik BN, Jacobs S, Westreich L, Mittelman S, Bausch JN (1996) Isolation and characterization of a monomethioninesulfoxide variant of interferon alpha-2b. Pharm Res 13:762–769

Glaring MA, Skryhan K, Kotting O, Zeeman SC, Blennow A (2012) Comprehensive survey of redox sensitive starch metabolising enzymes in *Arabidopsis thaliana*. Plant Physiol Biochem 58:89–97

Griffin TJ, Sherman J, Aebersold R (2001) Quantitative proteomics (ICAT™). Wiley, Hoboken

Groeger G, Quiney C, Cotter TG (2009) Hydrogen peroxide as a cell-survival signalling molecule. Antioxid Redox Signal 11:2655–2671

Gygi SP, Aebersold R (2000) Mass spectrometry and proteomics. Curr Opin Chem Biol 4:489–494

Gygi SP, Rist B, Gerber SA, Turecek F, Gelb MH, Aebersold R (1999) Quantitative analysis of complex protein mixtures using isotope-coded affinity tags. Nat Biotech 17:994–999

Harris TK, Turner GJ (2002) Structural basis of perturbed pKa values of catalytic groups in enzyme active sites. Int Union Biochem Mol Biol Life 53:85–98

Herbette S, Taconnat L, Hugouvieux V, Piette L, Magniette ML, Cuine S, Auroy P, Richaud P, Forestier C, Bourguignon J, Renou JP, Vavasseur A, Leonhardt N (2006) Genome-wide transcriptome profiling of the early cadmium response of Arabidopsis roots and shoots. Biochimie 88:1751–1765

Hu W, Huang C, Deng X, Zhou S, Chen L, Li Y, Wang C, Ma Z, Yuan Q, Wang Y, Cai R, Liang X, Yang G, He G (2013) TaASR1, a transcription factor gene in wheat, confers drought stress tolerance in transgenic tobacco. Plant Cell Environ 36:1449–1464

Huang XS, Liu JH, Chen XJ (2010) Overexpression of PtrABF gene, a bZIP transcription factor isolated from *Poncirus trifoliata*, enhances dehydration and drought tolerance in tobacco *via* scavenging ROS and modulating expression of stress-responsive genes. BMC Plant Biol 10:230

Huang GT, Ma SL, Bai LP, Zhang L, Ma H, Jia P, Liu J, Zhong M, Guo ZF (2012) Signal transduction during cold, salt, and drought stresses in plants. Mol Biol Rep 39:969–987

Huang XS, Wang W, Zhang Q, Liu JH (2013) A basic helix-loop-helix transcription factor, PtrbHLH, of *Poncirus trifoliata* confers cold tolerance and modulates peroxidase-mediated scavenging of hydrogen peroxide. Plant Physiol 162:1178–1194

Huang L, Yan H, Jiang X, Yin G, Zhang X, Qi X, Zhang Y, Yan Y, Ma X, Peng Y (2014) Identification of candidate reference genes in perennial ryegrass for quantitative RT-PCR under various abiotic stress conditions. Plos One 9:e93724. https://doi.org/10.1371/journal.pone.0093724

Hussain SS, Kayani MA, Amjad M (2011) Transcription factors as tools to engineer enhanced drought stress tolerance in plants. Biotechnol Prog 27:297–306

Jaffrey SR, Snyder SH (2001) The biotin switch method for the detection of S-nitrosylated proteins. Sci STKE 2001(86):pl1

Job C, Rajjou L, Lovigny Y, Belghazi M, Job D (2005) Patterns of protein oxidation in *Arabidopsis* seeds and during germination. Plant Physiol 138:790–802

Karuppanapandian T, Rhee SJ, Kim EJ, Han BK (2012) Proteomic analysis of differentially expressed proteins in the roots of Columbia-0 and Landsberg erecta ecotypes of *Arabidopsis thaliana* in response to aluminum-toxicity. Can J Plant Sci 92:1267–1282

Kawashima CG, Noji M, Nakamura M, Ogra Y, Suzuki KT, Saito K (2004) Heavy metal tolerance of transgenic tobacco plants over-expressing cysteine synthase. Biotechnol Lett 26:153–157

Ke Y, Han G, He H, Li J (2009) Differential regulation of proteins and phosphoproteins in rice under drought stress. Biochem Biophys Res Commun 379:133–138

Keller RJ, Halmes NC, Hinson JA, Pumford NR (1993) Immunochemical detection of oxidized proteins. Chem Res Toxicol 6:430–433

Khan M, Jan J, Karibe H, Komatsu S (2005) Identification of phosphoproteins regulated by gibberellins in rice leaf sheath. Plant Mol Biol 58:27–44

Khan NA, Singh S, Umar S (2008) Sulfur assimilation and abiotic stress in plants. Springer, Berlin/Heidelberg/Germany, p 372

Kim JB, Kang JY, Kim SY (2004) Over-expression of a transcription factor regulating ABA-responsive gene expression confers multiple stress tolerance. Plant Biotechnol J 2:459–466

Klatt P, Lamas S (2000) Regulation of protein function by S-glutathiolation in response to oxidative and nitrosative stress. Eur J Biochem 267:4928–4944

Knoefler D, Thamsen M, Koniczek M, Niemuth NJ, Diederich AK, Jakob U (2012) Quantitative in vivo redox sensors uncover oxidative stress as an early event in life. Mol Cell 47:767–776

Koller A, Washburn MP, Lange BM, Andon NL, Deciu C, Haynes PA, Hays L, Schieltz D, Ulaszek R, Wei J et al (2002) Proteomic survey of metabolic pathways in rice. Proc Natl Acad Sci U S A 99:11969–11974

Kopriva S, Mugford SG, Baraniecka P, Lee BR, Matthewman CA, Koprivova A (2012) Control of sulfur partitioning between primary and secondary metabolism in *Arabidopsis*. Front Plant Sci 3: 163

Kovtun Y, Chiu WL, Tena G, Sheen J (2000) Functional analysis of oxidative stress-activated mitogen-activated protein kinase cascade in plants. Proc Natl Acad Sci U S A 97:2940–2945

Kristensen BK, Askerlund P, Bykova NV, Egsgaard H, Moller IM (2004) Identification of oxidised proteins in the matrix of rice leaf mitochondria by immunoprecipitation and two-dimensional liquid chromatography tandem mass spectrometry. Phytochemistry 65:1839–1851

Kumar D, Datta R, Sinha R, Ghosh A, Chattopadhyay S (2014) Proteomic profiling of γ-ECS overexpressed transgenic *Nicotiana* in response to drought stress. Planta 9:e29246

Kwon SY, Jeong YJ, Lee HS, Kim JS, Cho KY, Allen RD, Kwak SS (2002) Enhanced tolerances of transgenic tobacco plants expressing both superoxide dismutase and ascorbate peroxidase in chloroplasts against methyl viologen-mediated oxidative stress. Plant Cell Environ 25:873–882

Kwon SY, Choi SM, Ahn YO, Lee HS, Lee HB, Park YM, Kwak SS (2003) Enhanced stress-tolerance of transgenic tobacco plants expressing a human dehydroascorbate reductase gene. J Plant Physiol 160:347–353

Lan P, Li W, Wen T-N, Shiau J-Y, Wu Y-C, Lin W, Schmidt W (2011) iTRAQ protein profile analysis of Arabidopsis roots reveals new aspects critical for iron homeostasis. Plant Physiol 155:821–834

Landry LG, Chapple CC, Last RL (1995) Arabidopsis mutants lacking phenolic sunscreens exhibit enhanced ultraviolet-B injury and oxidative damage. Plant Physiol 109:1159–1166

Lee SH, Ahsan N, Lee KW, Kim DH, Lee DG, Kwak SS, Kwon SY, Kim TH, Lee BH (2007) Simultaneous overexpression of both CuZn superoxide dismutase and ascorbate peroxidase in transgenic tall fescue plants confers increased tolerance to a wide range of abiotic stresses. J Plant Physiol 164:1626–1638

Leichert LI, Gehrke F, Gudiseva HV, Blackwell T, Ilbert M, Walker AK, Strahler JR, Andrews PC, Jakob U (2008) Proc Natl Acad Sci U S A 105:8197–8202

Levine RL, Stadtman ER (2001) Oxidative modification of proteins during aging. Exp Gerontol 36:1495–1502

Levine RL, Garland D, Oliver CN, Amici A, Climent I, Lenz A, Ahn B, Shaltiel S, Stadtman ER (1990) Determination of carbonyl content in oxidatively modified proteins. Methods Enzymol 186:464–478

Levine RL, Williams J, Stadtman ER, Shacter E (1994) Carbonyl assays for determination of oxidatively modified proteins. Methods Enzymol 233:346–357

Levine RL, Mosoni L, Berlett BS, Stadtman ER (1996) Methionine residues as endogenous antioxidants in proteins. Proc Natl Acad Sci U S A 93:15036–15040

Li Q, Fan CM, Zhang XM, Fu YF (2012) Validation of reference genes for real-time quantitative PCR normalization in soybean developmental and germinating seeds. Plant Cell Rep 31:1789–1798

Liang P, Pardee A (1992) Differential display of eukaryotic messenger RNA by means of the polymerase chain reaction. Science 257:967–971

Link AJ, Eng J, Schieltz DM, Carmack E, Mize GJ, Morris DR, Garvik BM (1999) Direct analysis of protein complexes using mass spectrometry. Nat Biotechnol 17:676–682

Liu JX, Bennett J (2011) Reversible and irreversible drought-induced changes in the anther proteome of rice (*Oryza sativa* L.) genotypes IR64 and moroberekan. Mol Plant 4:59–69

Liu GT, Ma L, Duan W, Wang BC, Li JH, Xu HG, Yan XQ, Yan BF, Li S, Wang L, Li J (2014) Differential proteomic analysis of grapevine leaves by iTRAQ reveals responses to heat stress and subsequent recovery. BMC Plant Biol 14:110

Lu H, Rate DN, Song JT, Greenberg JT (2003) ACD6, a novel ankyrin protein, is a regulator and an effector of salicylic acid signaling in the Arabidopsis defense response. Plant Cell 15:2408–2420

Lumbreras V, Vilela B, Irar S, Sole M, Capellades M, Valls M, Coca M, Pagès M (2010) MAPK phosphatase MKP2 mediates disease responses in Arabidopsis and functionally interacts with MPK3 and MPK6. Plant J 63:1017–1030

Ma Y, Szostkiewicz I, Korte A, Moes D, Yang Y, Christmann A, Grill E (2009) Regulators of PP2C phosphatase activity function as abscisic acid sensors. Science 324:1064–1068

Ma Y, Qin F, Tran LSP (2012) Contribution of genomics to gene discovery in plant abiotic stress responses. Mol Plant 5:1176–1178

MacCoss MJ, McDonald WH, Saraf A, Sȧdygov R, Clark JM, Tasto JJ, Gould KL (2002) Shotgun identification of protein modifications from protein complexes and lens tissue. Proc Natl Acad Sci U S A 99:7900–7905

Madian AG, Regnier FE (2010) Proteomic Identification of Carbonylated Proteins and Their Oxidation Sites. J Proteome Res 9:3766–3780

Maier KL, Lenz AG, Beck-Speier I, Costabel U (1995) Analysis of methionine sulfoxide in proteins. Methods Enzymol 251:455–461

Matsumura T, Tabayashi N, Kamagata Y, Souma C, Saruyama H (2002) Wheat catalase expressed in transgenic rice can improve tolerance against low temperature stress. Physiol Plant 116:317–327

Miles GP, Samuel MA, Ellis BE (2009) Suppression of MKK5 reduces ozone-induced signal transmission to both MPK3 and MPK6 and confers increased ozone sensitivity in Arabidopsis thaliana. Plant Signal Behav 4:687–692

Mishra NS, Tuteja R, Tuteja N (2006) Signaling through MAP kinase networks in plants. Arch Biochem Biophys 452:55–68

Moller IM, Jensen PE, Hansson A (2007) Oxidative modifications of cellular proteins. Annu Rev Plant Biol 58:459–481

Najami N, Janda T, Barriah W, Kayam G, Tal M, Guy M, Volokita M (2008) Ascorbate peroxidase gene family in tomato, its identification and characterization. Mol Gen Genomics 279:171–182

Nakagami H, Soukupova H, Schikora A, Zarsky V, Hirt H (2006) A mitogen-activated protein kinase kinase kinase mediates reactive oxygen species homeostasis in Arabidopsis. J Biol Chem 281:38697–38704

Nakamura A, Goto S (1996) Analysis of protein carbonyls with 2, 4-dinitrophenylhydrazine and its antibodies by immunoblot in two-dimensional gel electrophoresis. J Biochem 119:768–774

Narsai R, Ivanova A, Ng S, Whelan J (2010) Defining reference genes in Oryza sativa using organ, development, biotic and abiotic transcriptome datasets. BMC Plant Biol 10:56

Navascues J, Perez-Rontome C, Sanchez DH, Staudinger C (2012) Oxidative stress is a consequence, not a cause, of aluminum toxicity in the forage legume *Lotus corniculatus*. New Phytol 193:625–636

Nicot N, Hausman JF, Hoffmann L, Evers D (2005) Housekeeping gene selection for real-time RT-PCR normalization in potato during biotic and abiotic stress. J Exp Bot 56:2907–2914

Ning J, Li X, Hicks LM, Xiong L (2010) A Raf-like MAPKKK gene DSM1 mediates drought resistance through reactive oxygen species scavenging in rice. Plant Physiol 152:876–890

Noji M, Saito K (2007) Metabolic engineering of sulfur assimilation in plants. In: Verpoorte R et al (eds) Applications of plant metabolic engineering. Springer, Dordrecht, pp 297–309

Nolan T, Hands RE, Bustin SA (2006) Quantification of mRNA using real-time RT-PCR. Nat Protoc 1:1559–1582

Ortiz-Masia D, Perez-Amador MA, Carbonell J, Marcote MJ (2007) Diverse stress signals activate the C1 subgroup MAP kinases of Arabidopsis. FEBS Lett 581:1834–1840

Ouellette AJ, Anderson LB, Barry BA (1998) Amine binding and oxidation at the catalytic site for photosynthetic water oxidation. Proc Natl Acad Sci U S A 95:2204–2209

Park SY, Fung P, Nishimura N, Jensen DR, Fujii H, Zhao Y, Lumba S, Santiago J, Rodrigues A, Chow TF, Alfred SE, Bonetta D, Finkelstein R, Provart NJ, Desveaux D, Rodriguez PL, McCourt P, Zhu JK, Schroeder JI, Volkman BF, Cutler SR (2009) Abscisic acid inhibits Type 2C protein phosphatases *via* the PYR/PYL family of START proteins. Science 324:1068–1071

Parker R, Flowers TJ, Moore AL, Harpham NVJ (2006) An accurate and reproducible method for proteome profiling of the effects of salt stress in the rice leaf lamina. J Exp Bot 57(5):1109–1118

Pastore A, Piemonte F (2012) S-Glutathionylation signaling in cell biology: progress and prospects. Eur J Pharm Sci 46:279–292

Pimentel D, Haeussler DJ, Matsui R, Burgoyne JR, Cohen RA, Bachschmid MM (2012) Regulation of cell physiology and pathology by protein S-glutathionylation: lessons learned from the cardiovascular system. Antioxid Redox Signal 16:524–542

Pitzschke A (2012) Make Your Best-MYB transcription factors for improving abiotic stress tolerance in crops (Tuteja N, Gill SS, Tiburcio AF, Tuteja R (eds)). Wiley-VCH Verlag GmbH & Co. KGaA, Weinheim, pp 481–506

Pitzschke A (2013) From bench to barn, plant model research and its applications in agriculture. Adv Genet Eng. https://doi.org/10.4172/2169-0111.1000110

Prasad TK (1996) Mechanisms of chilling-induced oxidative stress injury and tolerance in developing maize seedlings, changes in antioxidant system, oxidation of proteins and lipids, and protease activities. Plant J 10:1017–1026

Prashanth S, Sadhasivam V, Parida A (2008) Over expression of cytosolic copper/zinc superoxide dismutase from a mangrove plant Avicennia marina in indica Rice var Pusa Basmati-1 confers abiotic stress tolerance. Transgenic Res 17:281–291

Rabilloud T, Chevallet M, Luche S, Leize-Wagner E (2005) Oxidative stress response, a proteomic view. Exp Rev Pro 2:949–956

Rausch T, Wachter A (2005) Sulfur metabolism, a versatile platform for launching defence operations. Trends Plant Sci 10:503–509

Reddy VY, Desorchers PE, Pizzo SV, Gonias SL, Sahakian JA, Levine RL, Weiss SJ (1994) Oxidative dissociation of human alpha 2-macroglobulin tetramers into dysfunctional dimmers. J Biol Chem 269:4683–4691

Rentel RC, Lecourieux D, Ouaked F, Usher SL, Petersen L, Okamoto H, Knight H, Peck SC, Grierson CS, Hirt H, Knight MR (2004) OXI1 kinase is necessary for oxidative burst-mediated signalling in Arabidopsis. Nature 427:858–861

Robinson CE, Kashavarzian A, Pasco DS, Frommel TO, Winshop DH, Holmes EW (1999) Determination of protein carbonyl groups by immunoblotting. Anal Biochem 266:48–57

Rodriguez MC, Petersen M, Mundy J (2010) Mitogen-activated protein kinase signaling in plants. Annu Rev Plant Biol 61:621–649

Roos G, Messens J (2011) Protein sulfenic acid formation: From cellular damage to redox regulation. Free Radic Biol Med 51:314–326

Roxas VP, Smith RK, Allen ER, Allen RD (1997) Overexpression of glutathione S-transferase/glutathione peroxidase enhances the growth of transgenic tobacco seedlings during stress. Nat Biotechnol 15:988–991

Saito K (2000) Regulation of sulfate transport and synthesis of sulfur-containing amino acids. Curr Plant Biol Opin 3:188–195

Salekdeh GH, Siopongco J, Wade LJ, Ghareyazie B, Bennett J (2002) Proteomic analysis of rice leaves during drought stress and recovery. Proteomics 2:1131–1145

Samelson LE (2001) Diagonal gel electrophoresis. Current protocols in immunology, Chapter 8, Unit 8.6.

Sarry JE, Kuhn L, Ducruix C, Lafaye A, Junot C, Hugouvieux V, Jourdain A, Bastien O, Fievet JB, Vailhen D, Amekraz B, Moulin C, Ezan E, Garin J, Bourguignon J (2006) The early responses of *Arabidopsis thaliana* cells to cadmium exposure explored by protein and metabolite profiling analyses. Proteomics 6:2180–2198

Sato Y, Masuta Y, Saito K, Murayama S, Ozawa K (2011) Enhanced chilling tolerance at the booting stage in rice by transgenic overexpression of the ascorbate peroxidase gene, OsAPXa. Plant Cell Rep 30:399–406

Schena M, Shalon D, Davis RW, Brown PO (1995) Quantitative monitoring of gene expression patterns with a complementary DNA microarray. Science 270:467–470

Schulze A, Downward J (2001) Navigating gene expression using microarrays–a technology review. Nat Cell Biol 3:E190–E195

Schurmann P, Buchanan BB (2008) The ferredoxin/thioredoxin system of oxygenic photosynthesis. Antioxid Redox Signal 10:1235–1273

Seki M, Ishida J, Narusaka M, Fujita M, Nanjo T, Umezawa T, Kamiya A, Nakajima M, Enju A, Sakurai T, Satou M, Akiyama K, Yamaguchi-Shinozaki K, Carninci P, Kawai J, Hayashizaki Y, Shinozaki K (2002) Monitoring the expression pattern of around 7,000 Arabidopsis genes under ABA treatments using a full-length cDNA microarray. Funct Integr Genomics 2:282–291

Sell S, Lindermayr C, Durner J (2008) Identification of S-nitrosylated proteins in plants. Methods Enzymol 440:283–293

Sethuraman M, McComb ME, Huang H, Huang S, Heibeck T, Costello CE, Cohen RA (2004) Isotope-coded affinity tag (ICAT) approach to redox proteomics, identification and quantitation of oxidant-sensitive cysteine thiols in complex protein mixtures. J Proteome Res 3:1228–1233

Shacter E, Williams JA, Lim M, Levine RL (1994) Differential susceptibility of plasma proteins to oxidative modification. Examination by Western blot immunoassay. Free Radic Biol Med 17:429–437

Shi S, Wang G, Wang Y, Zhang L, Zhang L (2005) Protective effect of nitric oxide against oxidative stress under ultraviolet-B radiation. Nitric Oxide 13:1–9

Sinha AK, Jaggi M, Raghuram B, Tuteja N (2011) Mitogen-activated protein kinase signaling in plants under abiotic stress. Plant Signal Behav 6:196–203

Smékalová V, Doskočilová A, Komis G, Šamaj J (2014) Crosstalk between secondary messengers, hormones and MAPK modules during abiotic stress signalling in plants. Biotechnol Adv 32:2–11

Smith MA, Sayre LM, Anderson VE, Harris PLR, Beal MF, Kowall N, Perry G (1998) Cytochemical demonstration of oxidative damage in Alzheimer disease by immunochemical enhancement of the carbonyl reaction with 2,4-dinitrophenylhydrazine. J Histochem Cytochem 46:731–735

Stadtman ER (1992) Protein oxidation and aging. Science 257(5074):1220–1224

Stadtman ER (2001) Protein oxidation in aging and age-related diseases. Ann N Y Acad Sci 928:22–38

Stadtman ER (2006) Protein oxidation and aging. Free Radic Res 40(12):1250–1258

Stadtman ER, Moskovitz J, Levine RL (2003) Oxidation of methionine residues of proteins: biological consequences. Antioxid Redox Signal 5(5):577–582

Stears RL, Martinsky T, Schena M (2003) Trends in microarray analysis. Nat Med 9:140–145

Strable J, Borsuk L, Nettleton D, Schnable PS, Irish EE (2008) Microarray analysis of vegetative phase change in maize. Plant J 56:1045–1057

Sullivan DM, Wehr NB, Fergusson MM, Levine RL, Finkel T (2000) Identification of oxidant-sensitive proteins, TNF-alpha induces protein glutathiolation. Biochemist 39:11121–11128

Takahashi H, Watanabe-Takahashi A, Smith F, Blake-Kalff M, Hawkesford M, Daito K (2000) The role of three functional sulfate transporters involved in uptake and translocation of sulphate in Arabidopsis thaliana. Plant J 23:171–182

Takesawa T, Ito M, Kanzaki H, Kameya N, Nakamura I (2002) Over-expression of zeta glutathione S-transferase in transgenic rice enhances germination and growth at low temperature. Mol Breed 9:93–101

Tanaka Y, Hibino T, Hayashi Y, Tanaka A, Kishitani S, Takabe T, Yokota S (1999) Salt tolerance of transgenic rice overexpressing yeast mitochondrial Mn-SOD in chloroplasts. Plant Sci 148:131–138

Taski-Ajdukovic K, Nagl N, Kovacev L, Curcic Z, Danojevic D (2012) Development and application of qRT-PCR for sugar beet gene expression analysis in response to *in vitro* induced water deficit. Electron J Biotechnol North America 1529:10

Teige M, Scheikl E, Eulgem T, Doczi R, Ichimura K, Shinozaki K, Dangl JL, Hirt H (2004) The MKK2 pathway mediates cold and salt stress signaling in Arabidopsis. Mol Cell 15:141–152

Udvardi MK, Kakar K, Wandrey M, Montanari O, Murray J, Andriankaja A, Zhang JY, Benedito V, Hofer JM, Chueng F, Town CD (2007) Legume transcription factors, global regulators of plant development and response to the environment. Plant Physiol 144:538–549

Vanguilder HD, Vrana KE, Freeman WN (2008) Twenty-five years of quantitative PCR for gene expression analysis. BioTechniques 44:619–626

Velculescu VE, Zhang L, Vogelstein B, Kinzler KW (1995) Serial analysis of gene expression. Science 270:484–487

Vera-Estrella R, Barkla BJ, Pantoja O (2014) Comparative 2D-DIGE analysis of salinity responsive microsomal proteins from leaves of salt-sensitive *Arabidopsis thaliana* and salt-tolerant *Thellungiella salsuginea*. Proteomics. https://doi.org/10.1002/pmic.201300123

Verslues PE, Lasky JR, Juenger TE, Liu TW, Kumar MN (2014) Genome-wide association mapping combined with reverse genetics identifies new effectors of low water potential-induced proline accumulation in Arabidopsis. Plant Physiol 164:144–159

Wan XY, Liu JY (2008) Comparative proteomics analysis reveals an intimate protein network provoked by hydrogen peroxide stress in rice seedling leaves. Mol Cell Proteomics 7(8):1469–1488

Wang Linqian DY, Zhiguang T (2010) iTRAQ labeling and biomarker discovery in comparative proteomic studies. Chem Life 30:135–140
Wang F-Z, Wang Q-B, Kwon S-Y, Kwak S-S, Su W-A (2005) Enhanced drought tolerance of transgenic rice plants expressing a pea manganese superoxide dismutase. J Plant Physiol 162:465–472. 51
Wang Z, Gerstein M, Snyder M (2009) RNA-Seq, a revolutionary tool for transcriptomics. Nat Rev Genet 10:57–63
Wang L, Li P, Brutnell TP (2010) Exploring plant transcriptomes using ultra high-throughput sequencing. Brief Funct Genomics 9:118–128
Wang C, Deng P, Chen L, Wang X, Ma H, Hu W, Yao N, Feng Y, Chai R, Yang G, He G (2013a) A wheat WRKY transcription factor TaWRKY10 confers tolerance to multiple abiotic stresses in transgenic tobacco. PloS One 8:e65120
Wang CY, Shen RF, Wang C, Wang W (2013b) Root protein profile changes induced by Al exposure in two rice cultivars differing in Al tolerance. J Proteome 78:281–293
Wang T, Hao R, Pan H, Cheng T, Zhang Q (2014a) Selection of suitable reference genes for quantitative real-time polymerase chain reaction in *Prunus mume* during flowering stages and under different abiotic stress conditions. J Am Soc Hortic Sci 139:113–122
Wang ZQ, Xu XY, Gong QQ, Xie C, Fan W, Yang JL, Lin QS, Zheng SJ (2014b) Root proteome of rice studied by iTRAQ provides integrated insight into aluminum stress tolerance mechanisms in plants. J Proteome 98:189–205
Washburn MP, Wolters D, Yates JR (2001) Large-scale analysis of the yeast proteome by multidimensional protein identification technology. Nat Biotech 19:242–247
Wehr NB, Levine RL (2013) Quantification of protein carbonylation. Methods Mol Biol 965:265–281
Wirtz M, Hell R (2003) Production of cysteine for bacterial and plant biotechnology, Application of cysteine feedback-insensitive isoforms of serine acetyltransferase. Amino Acids 24:195–203
Wolters DA, Washburn MP, Yates JR (2001) An automated multidimensional protein identification technology for shotgun analysis. Anal Chem 73:5683–5690
Wu WW, Baek GWSJ, Shen RF (2005) Comparative study of three proteomic quantitative methods, DIGE, clCAT, and iTRAQ, using 2D Gel- or LC- MALDI TOF/TOF. J Proteome Res 5:651–658
Xiong Y, Uys JD, Tew KD, Townsend DM (2011) S-glutathionylation: from molecular mechanisms to health outcomes. Antioxid Redox Signal 15:233–270
Yamaguchi-Shinozaki K, Shinozaki K (2006) Transcriptional regulatory networks in cellular responses and tolerance to dehydration and cold stresses. Annu Rev Plant Biol 57:781–803
Yan LJ (2009) Analysis of oxidative modification of proteins. Curr Protoc Protein Sci 14:144
Yan LJ, Forster MJ (2011) Chemical probes for analysis of carbonylated proteins. J Chromatogr B Analyt Technol Biomed Life Sci 879:1308–1315
Yan LJ, Sohal RS (1998) Mitochondrial adenine nucleotide translocase is modified oxidatively during aging. Proc Natl Acad Sci U S A 95:12896–12901
Yan LJ, Sohal RS (2000) Prevention of flight activity prolongs the life span of the housefly, Musca domestica, and attenuates the age-associated oxidative damage to specific mitochondrial proteins. Free Radic Biol Med 29:1143–1150
Yan LJ, Levine RL, Sohal RS (1997) Oxidative damage during aging targets mitochondrial aconitase. Proc Natl Acad Sci U S A 94:11168–11172
Yan LJ, Orr WC, Sohal RS (1998) Identification of oxidized proteins based on sodium dodecyl sulphate–polyacrylamide gel electrophoresis, immunochemical detection, isoelectric focusing, and microsequencing. Anal Biochem 263:67–71
Yan LJ, Christians ES, Liu L, Xiao X, Sohal RS, Benjamin IJ (2002) Mouse heat shock transcription factor 1 deficiency alters cardiac redox homeostasis and increases mitochondrial oxidative damage. EMBO J 21:5164–5172
Yan S, Tang Z, Su W, Sun W (2005) Proteomic analysis of salt stress-responsive proteins in rice root. Proteomics 5:235–244

Yang LM, Tian DG, Todd CD, Luo YM, Hu XY (2013) Comparative proteome analyses reveal that nitric oxide is an important signal molecule in the response of rice to aluminum toxicity. J Proteome Res 12:1316–1330

Yin L, Wang S, Eltayeb AE, Uddin MI, Yamamoto Y, Tsuji W, Takeuchi Y, Tanaka K (2010) Overexpression of dehydroascorbate reductase, but not monodehydroascorbate reductase, confers tolerance to aluminum stress in transgenic tobacco. Planta 231:609–621

Ying J, Clavreul N, Sethuraman M, Adachi T, Cohen RA (2007) Thiol oxidation in signaling and response to stress: detection and quantification of physiological and pathophysiological thiol modifications. Free Radic Biol Med 43:1099–1108

Youssefian S, Nakamura M, Orudgev E, Kondo N (2001) Increased cysteine biosynthesis capacity of transgenic tobacco overexpressing an O-acetylserine (thiol) lyase modifies plant responses to oxidative stress. Plant Physiol 126:1001–1011

Zhao F, Zhang H (2006) Salt and paraquat stress tolerance results from co-expression of the Suaeda salsa glutathione S-transferase and catalase in transgenic rice. Plant Cell Tiss Org Cult 86:349–358

Zhen Y, Qi JL, Wang SS, Su J (2007) Comparative proteome analysis of differentially expressed proteins induced by Al toxicity in soybean. Physiol Plant 131:542–554

Zheng M, Meng Y, Yang C, Zhou Z, Wang Y, Chen B (2014) Protein expression changes during cotton fiber elongation in response to drought stress and recovery. Int J Proteomics. https://doi.org/10.1155/2014/163962

Zhou S, Hu W, Deng X, Ma Z, Chen L, Huang C, Wang C, Wang J, He Y, Yang G, He G (2012) Overexpression of the wheat aquaporin gene, TaAQP7, enhances drought tolerance in transgenic tobacco. PloS One 7:e52439

Zhu YL, Pilon-Smits EAH, Jouanin L, Terry T (1999) Overexpression of glutathione synthetase in *Brassica juncea* enhances cadmium accumulation and tolerance. Plant Physiol 119:73–79

Zhu Y, Fu J, Zhang J, Liu T, Jia Z, Wang J, Jin Y, Lian Y, Wang M, Zheng J, Hou W, Wang G (2009) Genome-wide analysis of gene expression profiles during ear development of maize. Plant Mol Biol 70:63–77

Zieske LR (2006) A perspective on the use of iTRAQ reagent technology for protein complex and profiling studies. J Exp Bot 57:1501–1508

Zweier JL, Chen CA, Druhan LJ (2011) S-glutathionylation reshapes our understanding of endothelial nitric oxide synthase uncoupling and nitric oxide/reactive oxygen species-mediated signaling. Antioxid Redox Signal 14:1769–1775